MONDAY — Patterning and

1. 77 − 9 =

2. 9 × ___ = 27

3. Complete the following pattern:

    2, 3, 5, 8, ___, ___, ___

4. What is the rule for the following pattern?

    5, 8, 11, 14, 17, 20

5. Write the first three numbers for this pattern rule:

    start at 90, subtract 4

TUESDAY — Number Sense

1. Write 9545 in words.

2. Order these numbers from least to greatest.

    1.01, 0.11, 1.10, 0.01

3. Write 34 987 in expanded form.

4. Add 2.8 + 4.1

5. Write three hundred sixty-one dollars and fifteen cents in numbers.

Week 1

WEDNESDAY Geometry

1. Classify the following pair of lines.

   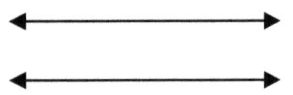

   A. intersecting
   B. parallel
   C. perpendicular

2. How many sides does a circle have?

3. Name this shape.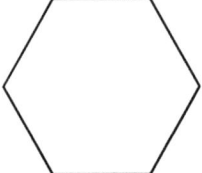

4. How many right angles does a square have?

5. How many faces does a cube have?

THURSDAY Measurement

1. 6 m = _____ mm

2. What time is 15:00 on the 12 hour clock?

3. How many ml does a regular can of pop hold?

4. What unit of measure would you use to find the mass of a person?
   A. mg
   B. g
   C. kg

5. What is the perimeter of a square that has 5 cm sides?

FRIDAY — Data Management

Laurel's class surveyed the teachers to see how they got to school everyday. Here are their results:

Car	Bus	Walk	Bike
‖‖‖‖ ‖‖‖‖	‖‖‖‖	‖‖‖‖	‖‖

1. How many teachers were surveyed? _____
2. How do most teachers come to school? _____
3. What is the range of the data? _____
4. What is the median of the data? _____
5. Draw a bar graph to represent the data above.

BRAIN STRETCH

Jonathon delivers 23 papers every day. His friend Marc delivers 17 papers every day.

a) How many papers do they deliver in a week?

b) How many more papers does Johnathon deliver in a week than Marc?

Week 1
© Chalkboard Publishing

MONDAY — Patterning and Algebra

1. Complete the pattern:

   12, 24, 36, ___, ___, ___

2. What is the rule for the following pattern?

   88, 77, 66, 55, 44, 33, 22, 11

3. What is the missing number?

   11 X ___ = 88

4. What will be the 20th shape in this pattern?

   □ ○ △ □ ○

5. Write the first three numbers for this pattern rule:

   start at 2, add 13

TUESDAY — Number Sense

1. Put the following numbers in order from least to greatest:

   21, 2.02, 2.11

2. Write the following in standard form:

   10 000 + 2000 + 7

3. Round this number to the nearest ten.

   349

4. Add $3.11 + $7.01

5. Multiply:

   3 x 10

WEDNESDAY Geometry

1. Name this shape:

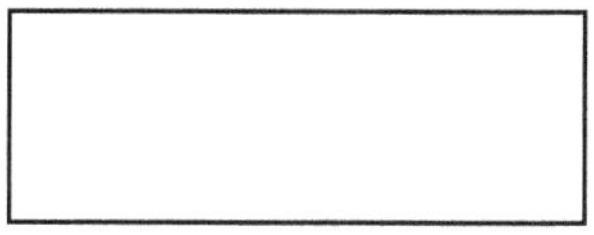

2. What is an obtuse angle?

3. How many vertices does a hexagon have?

4. How many angles does a triangle have?

5. What 3D figure can be made from these pieces?

THURSDAY Measurement

1. Find the perimeter of this shape:

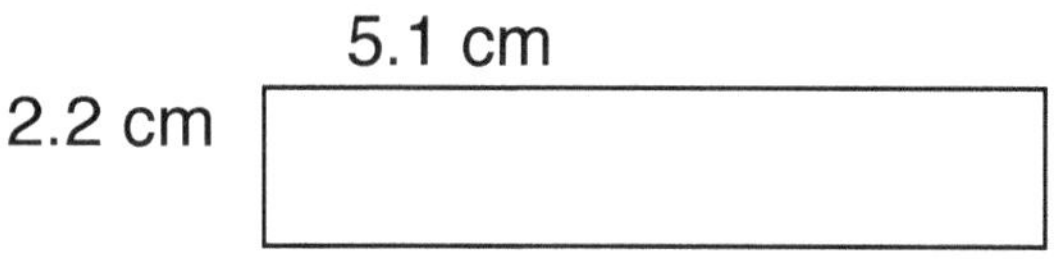

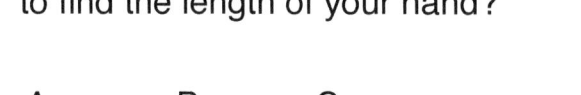
5.1 cm
2.2 cm

2. 200 cm=____dm

3. What unit of measurement would you use to find the length of your hand?

   A. mm   B. cm   C. m

4. What time is it?

5. The time is 10:34 a.m. What time will it be in 85 minutes?

Week 2

FRIDAY Data Management

Look at the spinner and answer the questions.

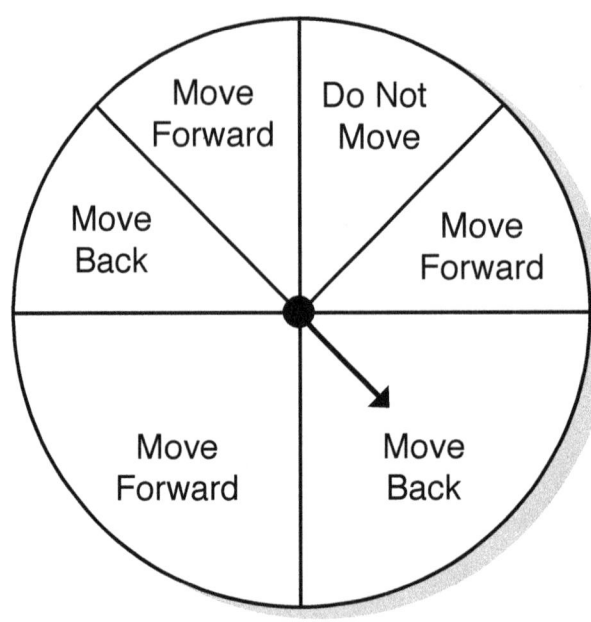

1. Is it more likely to land on Move Forward or Move Back? _____

2. What is the probability of landing on Move Forward? _____

3. What is the probability of landing on Move Back? _____

4. What is the probability of landing on Do Not Move? _____

BRAIN STRETCH

A car drove 50 km/hr for 2 hours then it went 90km/hr for 1 hour. Altogether, how far did the car travel?

MONDAY — Patterning and Algebra

1. 72 divided by 9 =

2. Write the first three numbers for this pattern rule:

    start a 67, subtract 4

3. Complete the pattern:

    101, 91, 81, ____, ____, ____

4. What is the rule for the following pattern?

    12, 14, 16, 18, 20, 22, 24

5. Is this a growing, shrinking or repeating pattern?

    11, 22, 11, 22, 11, 22, 11, 22

TUESDAY — Number Sense

1. Write 3432 in expanded form.

2. Write 2450 in words.

3. Multiply: 1.1 x 10

4. What is the largest number you can make with these digits:

    3 4 8 1

5. Maria bought six new tennis balls. Each ball was $0.89. How much did the six balls cost?

Week 3

WEDNESDAY — Geometry

1. What is a right angle?

2. How many sides does a pentagon have?

3. Name this shape:

   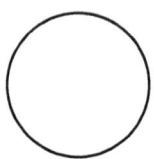

4. What 3D figure can be made from these pieces?

   A. pyramid
   B. cone
   C. cylinder

5. How many inside angles does a pentagon have?

THURSDAY — Measurement

1. What time is it 4 hours after 10:25 am?

2. Find the area of this rectangle.

   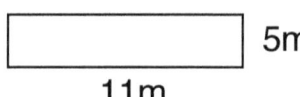
   5m
   11m

3. 12 mm = _____ cm

4. What unit of measurement would you use to find the mass of a paper clip?

   A. mg   B. g   C. kg

5. Which rectangle has the greater perimeter?

   A. 6 cm by 4 cm
   B. 6.5 cm by 2 cm

FRIDAY — Data Management

Mega Mart sold a lot of toys this summer. Here is a list of some toys they sold:

22 Pogo Sticks
46 Scooters
35 Skateboards
80 Skipping Ropes
42 Roller Blades
15 Hula Hoops

1. Create a bar graph using the data given.

2. What is the range of the data?_____

3. What was the most popular toy purchased?_____

4. What was the least popular toy purchased?_____

5. What is the mean amount of toys purchased?_____

BRAIN STRETCH

Nancy makes about $30 a week walking her neighbour's dog.

a) How much does she make in a year?

b) If Nancy puts half of her earnings in her savings account, how much does she save in one year?

Week 3

MONDAY — Patterning and Algebra

1. Write the first three numbers for this pattern rule:

    start at 900, add 15

2. Complete the pattern:

    5, 6, 8, 11, 15, ___ , ___ , ___

3. What is the following pattern rule?

    200, 225, 250, 275, 300

4. What will be the 12th number in this pattern:

    11, 22, 33, 11, 22, 33, 11

5. 56 + ___ = 71

TUESDAY — Number Sense

1. Spencer, Ben and Michael wanted to divide their bag of marbles. If they have 81 marbles, how many will each boy get?

2. Put the following numbers in order from greatest to least.

    909, 990, 0.99, 0.09

3. Round this number to the nearest hundred.

    54982

4. Subtract: 45.99 – 3.87

5. Which product is even?   A. 9 x 9   B. 2 x 10   C. 5 x 9

WEDNESDAY — Geometry

1. How many faces does a cone have?

2. How many inside angles does a rectangle have? What are they?

3. What is an acute angle?

4. What 3D figure can be made from this net?

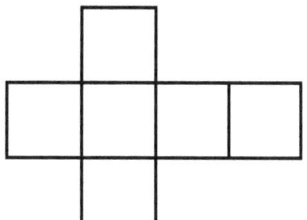

5. Name this shape: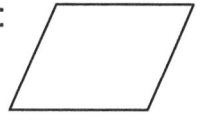

THURSDAY — Measurement

1. What unit of measurement would you use to find the height of your school?

2. How many metres are in 2500 km?

3. What time is it?

4. 1.1 m = _____ cm

5. If the perimeter of a hexagon is 60 cm, what is the length of each side?

Week 4

FRIDAY Data Management

Ashley had a bag full or marbles. She had 17 blue ones, 3 clear ones, 11 red ones, and 32 yellow ones.

1. How many marbles does Ashley have altogether? _____

2. What is the probability of randomly choosing a blue marble?

   _____

3. What is the probability of randomly choosing a clear marble?

   _____

5. What colour marble is most likely to be randomly picked from her bag?

   _____

6. What marble is least likely to be randomly chosen? _____

BRAIN STRETCH

Carrie's class collected school supplies for children who may not have them. They put 10 pencils, one ruler, 8 markers, 3 notebooks and 24 pencil crayons in each backpack.

a) How many pencils are there in 50 backpacks?

b) How many markers are there in 65 backpacks?

c) How many notebooks are there in 75 backpacks?

d) How many pencil crayons are there in 100 backpacks?

MONDAY — Patterning and Algebra

1. What is the following pattern rule?

    70, 65, 60, 55, 50, 45, 40

2. 108 = ____ - 21

3. Write the first three numbers for this pattern rule:

    start at 4, add 25

4. Complete the pattern:

    6, 12, 24, 48, __, __, __

5. What will be the 28th figure in this pattern?

TUESDAY — Number Sense

1. Add:

    234.1 + 29.6

2. Divide: 819 by 9

3. Write 2052 in words.

4. Write the following in standard form:

    7000 + 200 + 9 + 0.6

5. Sophie bought a t-shirt for $16.99 and a pair of sunglasses for $24.99. How much was her bill?

Week 5

WEDNESDAY — Geometry

1. How many sets of parallel lines does a square have?

2. Name this shape: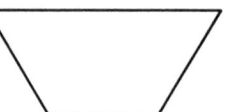

3. How many faces does a cylinder have?

4. Draw a straight angle.

5. What 3D figure can be made from this net?

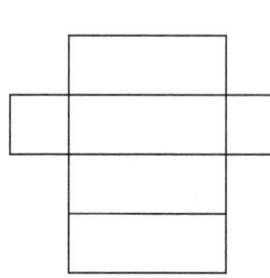

THURSDAY — Measurement

1. What time will it be 4 and a half hours after 3:15 pm?

2. Draw a triangle with a perimeter of 9 cm.

3. Donte's backyard measures 10 m by 18 m. What is the area?

4. If Donte wants to put a fence around his yard, how much fencing should he buy?

5. 80 dm = _____ mm

FRIDAY — Data Management

This graph shows the number of books borrowed at the Fairfield Elementary School library by some of its students.

The Number of Books Borrowed by Students

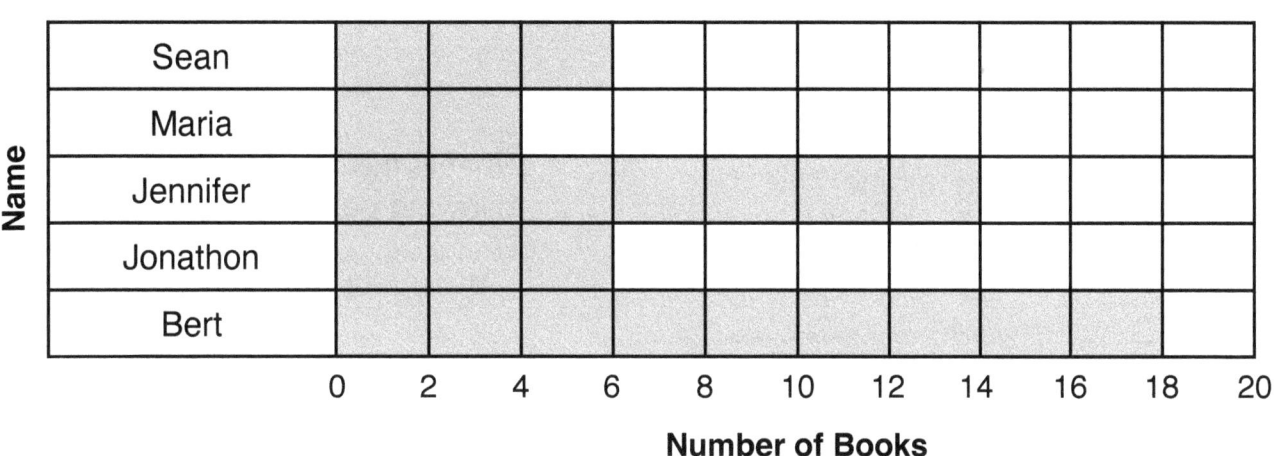

1. Who borrowed the most books at Fairfield Elementary? _____

2. Who borrowed the least number of books? _____

3. What is the range of the data? _____

4. What is the median amount? _____

5. What is the mean amount? _____

BRAIN STRETCH

a) A 5 kg bag of oranges sells for $7.50. What is the cost per kilogram?

b) How many 5 kg bags can you buy with $40?

Week 5

MONDAY — Patterning and Algebra

1. Is this a growing, shrinking or repeating pattern?

    2, 7, 12, 17, 22, 27, 32, 37

2. Write the first three numbers for this pattern rule:

    start at 222, subtract 18

3. What is the missing number?

    36 ÷ ____ = 6

4. What is the rule for the following pattern?

    45, 48, 51, 54, 57, 60

5. Complete the pattern:

    25, 75, 125, 175, ___, ___, ___

TUESDAY — Number Sense

1. How many thousands in 84 983?

2. Write 12.78 in expanded form.

3. Put the following numbers in order from least to greatest.

    2.2, 0.22, 20.2

4. Subtract: 2345 - 495

5. Multiply: 1.3 x 10

WEDNESDAY — Geometry

1. Draw a pair of parallel lines.

2. What is a reflex angle?

3. What 3D figure can be made from this net?

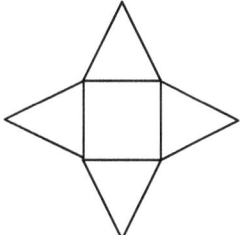

4. Name this shape.

5. How many vertices does a cone have?

THURSDAY — Measurement

1. 330 cm = _____ mm

2. What unit of measurement would you use to find the mass of an elephant?

   A. g   B. t   C. kg

3. How many minutes in 9 hours?

4. Find the area of this square.

   4 cm □

5. What time is it?

FRIDAY — Data Management

Mrs. Monroe took a survey of what foods her students preferred to buy from the cafeteria. Here are the results.

	Number of Votes
French Fries	25
Hamburgers	17
Patties	12
Sandwiches	15
Salads	2
Soups	5

1. What was the most popular choice? _____
2. What was the least popular choice? _____
3. What is the range of the data collected? _____
4. What is the median? _____
5. What is the mean? _____
6. How many students were surveyed? _____

BRAIN STRETCH

A movie ticket costs $9.50 per person. Popcorn is $3.25 and a drink is $2.35.

a) How much money does it cost to see a movie, eat popcorn and have a drink?

b) What would the change be from a twenty dollar bill?

MONDAY — Patterning and Algebra

1. Complete the pattern:

    14, 17, 20, 23, ___, ___, ___

2. What is the missing number?

    70 - ___ = 43

3. What will be the 15th number in this pattern?

    5, 10, 15, 20, 25

4. Write the first three numbers for this pattern rule:

    start at 7, multiply by 2

6. What is the rule for the following pattern?

    11, 22, 11, 22, 11, 22

TUESDAY — Number Sense

1. $74.76
    + $25.92
    _____

2. Write the smallest number possible with these digits:

    5   2   8   3

3. Chris had to give half of his sticker collection to his brother. He had 384 stickers. How many does he have left?

4. Write 1499 in words.

5. Round this number to the nearest ten.

    3986

Week 7   © Chalkboard Publishing

WEDNESDAY — Geometry

1. What is the name of this 3D figure?

   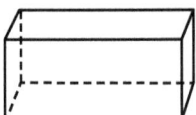

2. How many vertices does a cylinder have?

   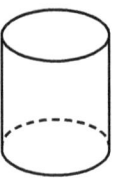

3. Name this shape:

   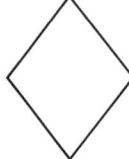

4. What does one face of a cube look like?

5. Draw intersecting lines.

THURSDAY — Measurement

1. What is the length of each side of an equilateral triangle with a perimeter of 15 m?

2. 66 mm=_____m

3. What unit of measurement would you use to find the distance across the Atlantic Ocean?

4. Which is the longest?

   A. 12 cm
   B. 12 dm
   C. 12 m

5. Perimeter= _____ units

   Area= _____ square units

   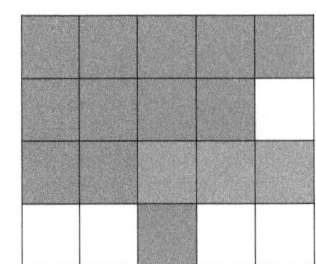

FRIDAY Data Management

Montrose Community Centre has a lot of after school sports clubs. 34 students play floor hockey, 58 play soccer, 31 play badminton, 49 play basketball and 11 play chess.

1. Make a tally chart of this data:

Floor hockey	Soccer	Badminton	Basketball	Chess

2. How many students participate in after school clubs at the community centre?
   _____

3. What is the most popular club? _____

4. What is the least popular club? _____

5. What is the mean of the data? _____

6. What is the range of the data? _____

BRAIN STRETCH

State the place value of the underlined digit in each number:

a. 31 <u>4</u>90 _____   b. 87 4<u>5</u>1 _____

c. 6<u>4</u>38 _____   d 2 45<u>6</u> _____

e. 5<u>7</u> 239 _____   f. <u>2</u>9 023 _____

Week 7

MONDAY — Patterning and Algebra

1. Write the first three numbers for this pattern rule:

    start at 245, add 11

2. What is the following pattern rule?

    100, 200, 300, 400, 500

3. 66 + _____ = 91

4. Is this a growing, shrinking or repeating pattern?

    100, 90, 80, 70, 60, 50

5. Extend the pattern.

    80, 91, 102, 113, ____, ____, ____

TUESDAY — Number Sense

1. Which number is 10 000 less than 22 182

    A. 21 182   B. 12 182   C. 21 000

2. Add: 56 + 6734

3. Mega Mart sells about 45 4L bags of milk each day. How many litres do they sell each week?

4. Put the following numbers in order from greatest to least.

    7.07, 7.0, 0.77, 77.7

5. Multiply: 23 x 8

WEDNESDAY — Geometry

1. What 3D figure could be made from these pieces?

2. Name this 3D figure.

   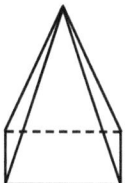

3. Which figure shows a line of symmetry?

   A.    B.    C.

4. Look at the shapes. Choose flip, slide or turn.

    →

5. Classify the following triangle.

   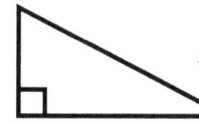   A. right   B. obtuse   C. equilateral

THURSDAY — Measurement

1. What is perimeter?

2. What time is it?

3. 1 km = _____ dm

4. What unit of measurement would you use to find the mass of a grain of sand?

5. Larry's stride is about 70 cm. It took him 8.5 strides to measure the length of his pool. About how long is the pool?

Week 8

© Chalkboard Publishing

FRIDAY — Data Management

Carlos wanted to design a spinner for his new game. He wanted to have a spinner with ¼ probability to land on red, ⅛ probability to land on green, ⅛ probability to land on yellow and ½ probability to land on blue.

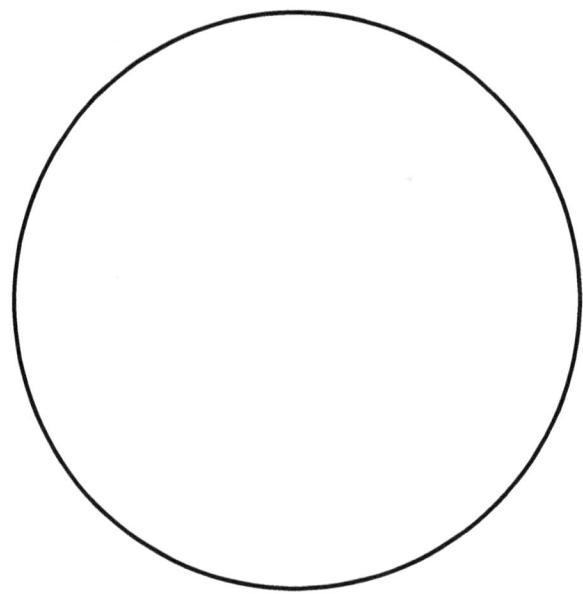

1. In the circle above, create a spinner like the one Carlos wanted.

2. What colour are you most likely to spin? _____

3. What colours are you equally likely to spin? _____

4. What colours are you least likely to spin? _____

BRAIN STRETCH

Marie's mom needs to buy 75 dinner rolls. They are sold 12 per bag.

a) How many bags does she need to buy?

b) If one dozen buns cost $1.99, how much will she need to spend?

MONDAY — Patterning and Algebra

1. What is the rule for the following pattern?

    20, 24, 28, 32, 36, 40, 44

2. Write the first three numbers for this pattern rule:

    start at 20, add 13

3. Extend the pattern.

    225, 220, 214, 207, ____, ____

4. What will be the 30th shape in this pattern?

5. 200 – 56 = X   X = _____

TUESDAY — Number Sense

1. Multiply 13.4 x 100

2. Round this number to the nearest hundred.

    8749

3. Write 698.2 in expanded form.

4. Last year, the Bulls scored 210 points. This season, they have scored 178. How many fewer points did they score this year?

5. Write 2250 in words.

Week 9

WEDNESDAY — Geometry

1. How many vertices does a rectangular prism have?

   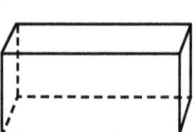

2. Name this shape.

   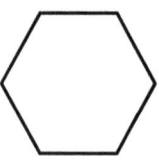

3. How many parallel lines does a triangle have?

4. Draw a trapezoid.

5. Which pair of shapes look congruent?

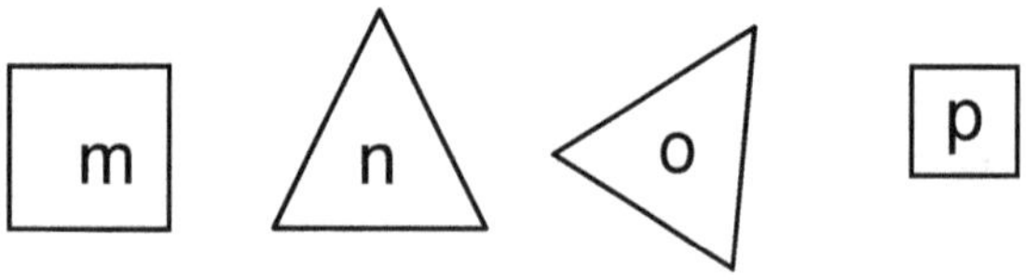

     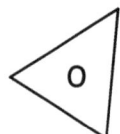

THURSDAY — Measurement

1. What unit of measurement would you use to find the width of a textbook?

2. What is the formula to calculate the area of a rectangle?

3. What unit of measure would you use to time how long it takes you to shower?

4. 990 mm = _____ dm

5. The time is 12:34 a.m. What time will it be in 75 minutes?

FRIDAY — Data Management

Nine planes leave City Centre Airport every hour. This graph shows the number of people departing today at noon.

City Centre Departures, 12 noon

City	0	20	40	60	80	100	120	140	160	180
Victoria	■	■	■	■	■	■				
London	■	■								
Halifax	■	■	■							
Ottawa	■	■	■	■						
Winnipeg	■	■	■	■						
North Bay	■	■	■							
Montreal	■	■	■	■	■	■	■	■	■	

Number of People

1. What flight has the most people on it? _____

2. What flights have the same number of passengers? _____

3. What is the range of the data? _____

4. What is the median number of people on the flights? _____

5. Find the mean amount of people on the flights. _____

BRAIN STRETCH

How many weeks are there in 10 years? In 100 years? In 1000 years?

MONDAY — Patterning and Algebra

1. a + b = 20,

    a=_____, b=_____

2. Complete the pattern:

    15, 21, 27, 33, __, __, __

3. What is the rule for the following pattern?

    130, 123, 116, 109, 102

4. Write the first three numbers for this pattern rule:

    start at 100, subtract 7

5. Is this a growing, shrinking or repeating pattern?

    12, 13, 14, 12, 13, 14

TUESDAY — Number Sense

1. Multiply: 78 x 7

2. Write six hundred forty-nine and two tenths in numbers.

3. Megan needs 4200 sheets of paper. How many packs should she buy if each pack has 500 sheets?

4. Divide: 100 by 4

5. Put the following numbers in order from least to greatest.

    55, 5.5, 5.55, 0.55

WEDNESDAY — Geometry

1. Draw a right angle.

2. How many edges?
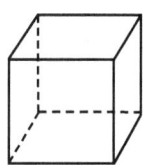

3. How many of the following pairs of line intersect?
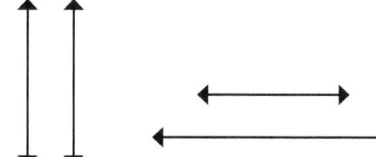

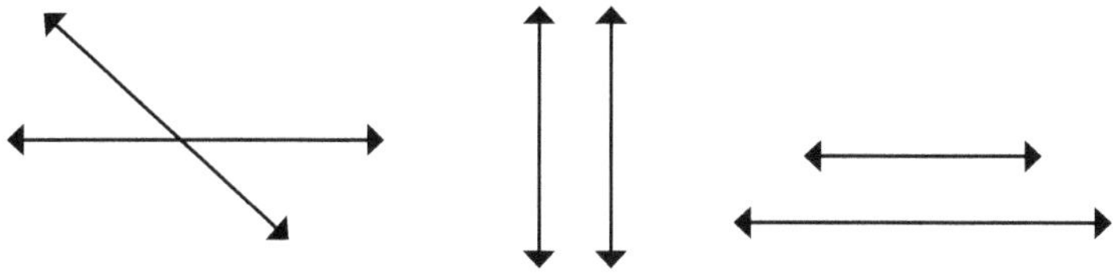

4. Name this 3D figure.
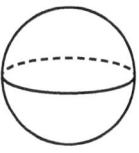

5. How many parallel lines does a hexagon have?

THURSDAY — Measurement

1. Katelyn's bedroom is 2m by 4m. What is the area of her room? What is the perimeter of her room?

2. What unit of measurement would you use to find the mass of a bag of potato chips?

   A. kg   B. g   C. mg

3. Find the perimeter of the trapezoid.
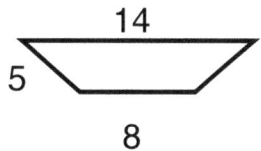

4. What time is it?

5. 50 cm = _____ m

Week 10

FRIDAY — Data Management

Laura's mom planned a scavenger hunt for her birthday party. She hid fifty plastic containers filled with toys. 6 containers had stuffed animals, 10 had puzzles, 8 had books, 20 had stickers and 6 had jacks.

1. What is the probability of getting a puzzle? _____

2. What it he probability of getting jacks? _____

3. What are you most likely to get? _____

4. What are you least likely to get? _____

5. Which of the prizes would you like to get? _____

BRAIN STRETCH

A box of dog biscuits has 18 biscuits. A box costs $2.88.

a) How much does 1 biscuit cost?

b) How many dog biscuits are in five boxes?

c) How much would five boxes cost?

d) If you paid for five boxes with a $20 bill, how much change would you get back?

MONDAY — Patterning and Algebra

1. Write the first three numbers for this pattern rule:

   start at 2, multiply by 2

2. $17 + C = 41$

   $C =$

3. Find the 20th number in this pattern:

   79, 77, 75, 73, 71, 69

4. Complete the pattern:

   2, 15, 28, 41, ___, ___, ___

5. What is the rule for the following pattern?

   400, 410, 420, 430, 440

TUESDAY — Number Sense

1. Write 9 901 in words.

2. Write the following in standard form:

   $80000 + 900 + 30 + 0.8$

3. Divide: 456 by 4

4. Compare using: $<$, $>$ or $=$.

   6487 ☐ 6487

5. David gave a $20 bill to pay for his $11.78 bill at the convenience store. How much change will he get back?

Week 11 © Chalkboard Publishing

WEDNESDAY — Geometry

1. Name this 3D figure.

2. How many faces does a sphere have?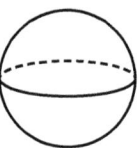

3. Draw an acute angle.

4. Complete: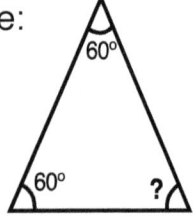

   measure of angle _____

   type of triangle _____

5. Which 3D figure has no edges?   A. 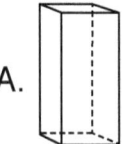   B. 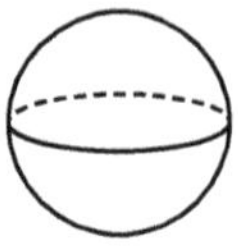 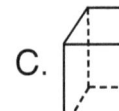   C.

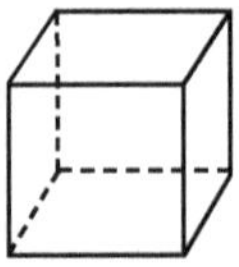

THURSDAY — Measurement

1. 300 dm = ____ mm

2. What time is it?

3. What unit of measurement would you use to find the mass of a baseball bat?

4. Draw a triangle with a perimeter of 12 cm.

5. How many kilograms in 800 grams?

FRIDAY — Data Management

Liam took a survey of his teachers and classmates and their favourite kind of television shows. He found that 24 people like to watch soap operas, 20 like sports, 32 like comedies, 12 like dramas, 41 like cartoons and 6 like the news.

Favourite T.V. Shows	Number of People
Soap Opera	
Sports	
Comedy	
Drama	
Cartoon	
News	

 = 2 people

1. Use the data given to make a pictograph in the chart above. Draw one T.V. set for every two people.

2. What is the mean? _____

3. What is the most favourite kind of T.V. show? _____

4. What is the range of the data? _____

5. What is the least favourite kind of T.V. show? _____

BRAIN STRETCH

Everyday a bus travels from Claireville to Milltown. Each day on average, 35 people travel on the bus.

 a) How many riders go on the trip in a week?

 b) 18 weeks?

 c) 1 year?

Week 11

MONDAY — Patterning and Algebra

1. Complete the pattern:

   1, 4, 16, 64 ,___, ___, ___

2. What is the rule for the following pattern?

   700, 689, 678, 667, 656

3. $6 \times B = 54$

   B=_____

4. Is this a growing, shrinking or repeating pattern?

   700, 698, 696, 694, 692

5. Write the first three numbers for this pattern rule:

   start at 75, add 15

TUESDAY — Number Sense

1. Put the following numbers in order from greatest to least.

   2.23, 3.32, 2.32, 3.22

2. Multiply: 45×15

3. Madelyn spends about 1.5 hours playing the piano every week. How much does she play in 10 weeks?

4. Write 7644.3 in expanded form.

5. Multiply: 4.5×10

WEDNESDAY — Geometry

1. How many lines of symmetry does this letter have?

   K

2. How many sides does a quadrilateral have?

3. Name this shape:

   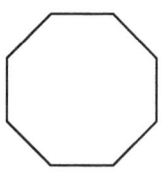

4. What 3D figure could be made from these pieces?

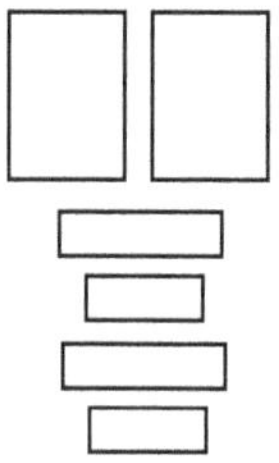

   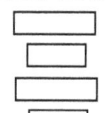

   A. cylinder

   B. rectangular prism

   C. pyramid prism

5. Draw an obtuse angle.

THURSDAY — Measurement

1. Find the area of this rectangle.

   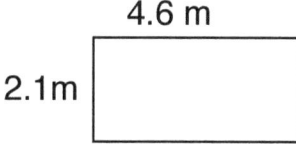

2. What is the perimeter of this pentagon?

   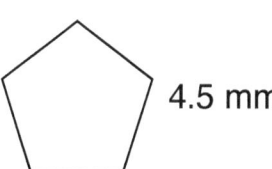

   4.5 mm

3. 44 dm = _____ km

4. What unit of measurement would you use to find the length of a fly's wing?

5. How many minutes are there in 12 hours?

Week 12

FRIDAY Data Management

Here is a pie graph to show students' favourite fruits.

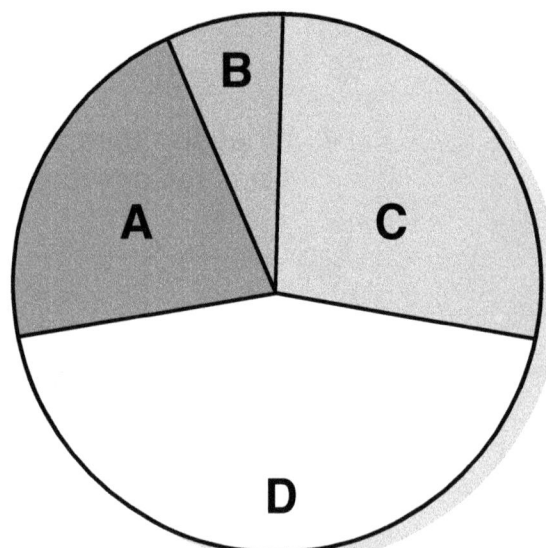

Students' Favourite Fruits

A - Apples
B - Bananas
C - Grapes
D - Strawberries

1. Which fruit was picked the most? _____

2. Which fruit was picked the least? _____

3. Which two fruits were picked about the same? _____

BRAIN STRETCH

What is the value of the money?

MONDAY — Patterning and Algebra

1. Is this a growing, shrinking or repeating pattern?

    89, 89, 89, 89, 89, 89

2. Write the first three numbers for this pattern rule:

    start at 61, add 8

3. Extend the pattern.

    500, 440, 380, 320, ____, ____, ____

4. What should replace the _____ to make the following equation true?

    5 × 12 = 600 _____ 10

    A. + B. - C. ÷

5. What is the rule for the following pattern:

    309, 306, 303, 300, 297

TUESDAY — Number Sense

1. Round this number to the nearest tenth.

    23.19

2. Write 78 215 in words.

3. Which fraction below is equivalent to $\frac{1}{2}$?

    $\frac{2}{3}$ $\frac{4}{7}$ $\frac{4}{8}$

4. Karen can skip about 45 times per minute. How many times could she skip in 5 minutes?

5. Divide: 122 by 2

Week 13

WEDNESDAY — Geometry

1. Which 3D figure does this object look like?

   _____

2. A rhombus is a polygon with how many sides?

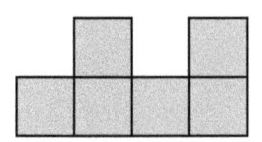

3. Is this net for a cube possible?

4. An angle of 120° is called:

   A. obtuse   B. right   C. acute

5. This is a _____.

   How many?   faces _____   edges _____   vertices _____

THURSDAY — Measurement

1. Joe's field measures 4.8 km long and 2.3 km wide. How much fencing will he need in order to surround his property?

2. How many decades in 40 years?

3. What time is it?

4. 163 cm = _____ m

5. Each side of a square measures 4 cm. What would the perimeter be if the sides were to increase by 2 cm each?

FRIDAY — Data Management

Aidan has 120 marbles. Half of the marbles are blue, ¼ are yellow, ⅛ are white and ⅛ are red.

1. How many marbles are blue? _____

2. How many marbles are yellow? _____

3. How many are white or red? _____

4. What colour do you have the greatest probability of randomly selecting? _____

5. What colour marble do you have the least probability of randomly selecting? _____

BRAIN STRETCH

Yummy Ice Cream Parlor sells single scoops for $1.25 and double scoops for $1.90.

a) How much will it cost to buy two single scoops and two double scoops?

b) How much change will you get back from a $20 bill?

c) What is the least number of bills and coins you can use to make the change?

Week 13

MONDAY — Patterning and Algebra

1. Complete the pattern:

   69, 76, 83, 90, ___, ___, ___

2. What is the rule for the following pattern?

   557, 559, 561, 563, 565

3. What will be the 34th number in this pattern?

   543, 540, 537, 534, 531

4. Write the first three numbers for this pattern rule:

   start at 89, add 20

5. 330 – 52 = ___

TUESDAY — Number Sense

1. Multiply: 3.4 by 100

2. Put the following numbers in order from least to greatest.

   $\frac{1}{2}$ $\frac{1}{3}$ $\frac{1}{4}$

3. Write the following in standard form:

   60000 + 40 + 3

4. How many dimes are in 10 toonies?

5. Add: 45.67 + 34.22

WEDNESDAY — Geometry

1. Classify the following triangle.

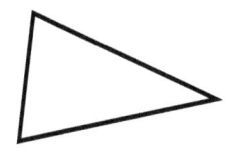

   A. scalene   B. isosceles   C. equilateral

2. How many faces?

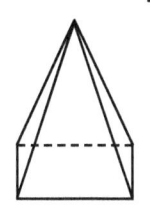

   A. 5   B. 4   C. 12

3. How many lines of symmetry?

   W

4. Draw a reflex angle.

5. Classify the angle:

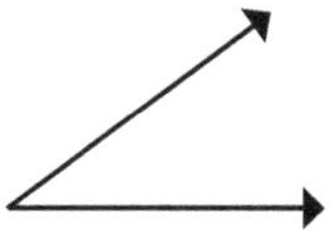

THURSDAY — Measurement

1. What unit of measure would you use to measure the capacity of a juice carton?

2. 90 km = _____ m

3. 27.1 m = _____ cm

4. What time is it?

5. John's desk surface is about 40 cm long and 40 cm wide. What is the area of the desk top?

Week 14

FRIDAY — Data Management

Here is a bar graph of how much money the grade 5 class fund raised for a class trip.

Trip Fund Raising

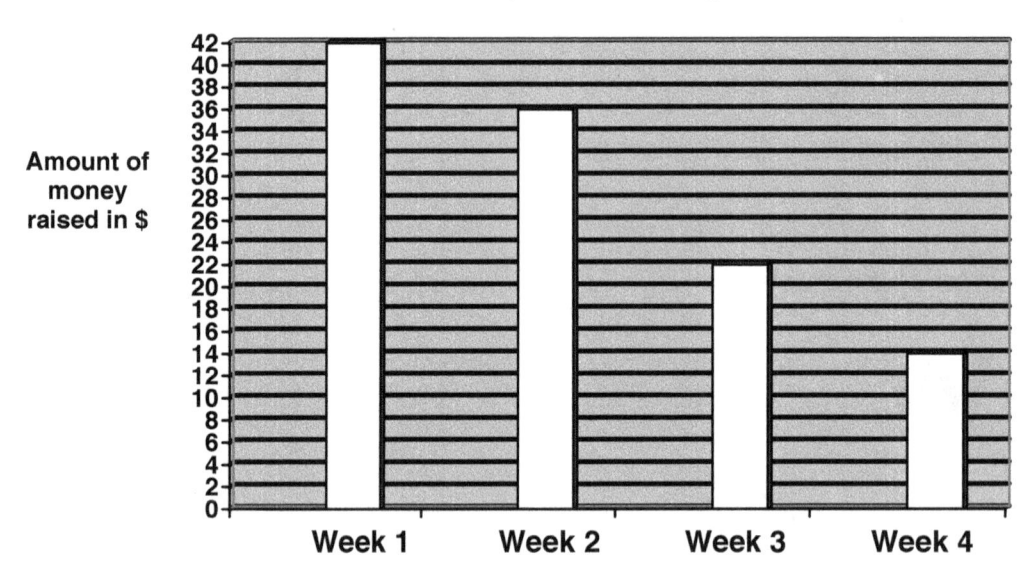

(Amount of money raised in $)
- Week 1: 42
- Week 2: 36
- Week 3: 22
- Week 4: 14

1. What week was the most money raised? _____

2. How much money was raised in week 3? _____

3. How much money was raised altogether? _____

4. How much more money was raised in week 1 than in week 4? _____

BRAIN STRETCH

a) How many pennies are in ten dollars?

b) How many pennies are in one hundred dollars?

c) How many pennies are in one thousand dollars?

MONDAY — Patterning and Algebra

1. Write the first three numbers for this pattern rule:

    start at 1000, subtract 50

2. Is this a growing, shrinking or repeating pattern?

    A, D, A, J, A, D, A, J, A, D, A, J

3. Complete the pattern:

    1, 10, 21, 34, 49, ____, ____, ____

4. What should replace the _____ to make the following equation true?

    5 × 11 = 60 _____ 5

    A. +   B. −   C. ÷

5. What is the rule for the following pattern?

    999, 1002, 1005, 1008, 1011

TUESDAY — Number Sense

1. Divide 7.9 by 10:

2.
    $$\begin{array}{r} 37 \\ \times\ 86 \\ \hline \end{array}$$

3. Divide:

    $2\overline{)4386}$

4. Write 1009.88 in expanded form.

5. Which of the following fractions is equivalent to $\frac{1}{4}$?

    $\frac{2}{8}$ $\frac{2}{3}$ $\frac{5}{10}$

Week 15

WEDNESDAY — Geometry

1. Look at the shapes. Choose flip, slide or turn.

2. Name two quadrilaterals.

3. How many lines of symmetry does this letter have?

   X

4. Classify the following pair of lines.

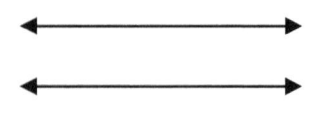

   A. intersecting   B. parallel   C. perpendicular

5. Complete: (triangle with 140°, 20°, ?)

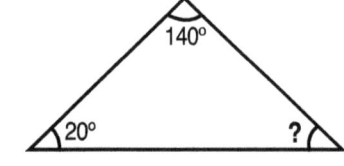

   measure of angle _____

   type of triangle _____

THURSDAY — Measurement

1. 350 km = _____ m

2. How many months in 5 years?

3. Mary's garden has a 23 m perimeter. If fencing costs $4 per meter, how much will a garden fence cost?

4. David lives 950 000 cm from his school. How many metres is that?

5. What time is it?

FRIDAY — Data Management

Mrs. Martin's class tossed a styrofoam cup and came up with these results:

Lands on Top	Lands on Bottom	Lands on Side

1. How many times did they toss the cup altogether? _____
2. How many times did it land on top? _____
3. How many times did it land on its bottom? _____
4. How many times did it land on its side? _____
5. Which outcome was most likely? _____
6. Write a fraction to show how many times it landed on top. _____
7. Write a fraction to show how many times it landed on its bottom. _____

BRAIN STRETCH

Find the product for each of the following:

a) 5.80 X 10 =

b) 4.05 X 1000 =

c) 216.35 X 100 =

d) 22.09 X 1000 =

e) 0.52 X 10 =

f) 0.79 x 100 =

MONDAY — Patterning and Algebra

1. What will be the 14th animal in the pattern?

2. Complete the pattern:

   99, 94, 89, 84, ___, ___

3. Write the first three numbers for this pattern rule:

   start at 600, add 17

4. What is the rule for the following pattern?

   246, 241, 236, 231, 226, 221

5. $46 \div b = 23$

   b=_____

TUESDAY — Number Sense

1. Subtract: 110 - 78.1

2. Multiply: 1.45 x 1000

3. Put the following in order from greatest to least:

   1 1/2, 1 1/4, 1 3/4

4. Round this number to the nearest tenth.

   6900.65

5. Sunita reads about 5 books a month. How many books does she read in a year?

WEDNESDAY — Geometry

1. What kind of triangle is this?

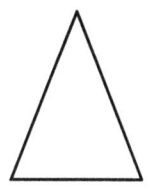

2. Draw the reflection of this shape.

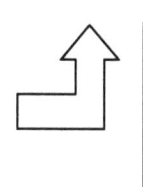

3. Draw a 90 degree angle.

4. Name a polygon with 3 sides.

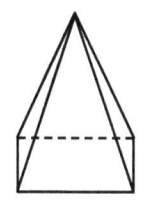

5. This is a _____.

How many?   faces _____   edges _____   vertices _____

THURSDAY — Measurement

1. What time is it?

2. Draw a triangle with an area of 10cm^2.

3. 125 mm = _____ cm

4. What unit of measurement would you use to find the length of your arm?

5. How would you find the perimeter of your classroom?

Week 16

FRIDAY — Data Management

The Mill Valley Hockey Association published the final results for this season.

Team	Overall Points
Tigers	23
Sharks	16
Mustangs	21
Hornets	18
Cobras	24
Scorpions	34
Barracudas	20

1. What is the range of the data? _____

2. What is the median number of points? _____

3. What is the mean number of points? _____

4. How many more points did the Cobras score than the Sharks? _____

5. How many fewer points did the Hornets have than the Scorpians? _____

BRAIN STRETCH

A tub of ice cream has 4L in it.

a) If each person at the Smith family BBQ gets a 200 ml scoop, how many people can be served by one tub?

b) If there are 35 people at the BBQ, how many tubs of ice cream should they purchase?

MONDAY — Patterning and Algebra

1. 4 x a = 52

    a = _____

2. What is the rule for the following pattern?

    50, 100, 200, 400, 800

3. Is this a growing, shrinking or repeating pattern?

    65, 64, 62, 59, 55, 50

4. Complete the following pattern:

    100, 89, 78, 67, ___, ___

5. Which pair of numbers best completes the equation? ☐ X 10 = ☐

    A. 120 and 1200   B. 1.2 and 120   C. 0.12 and 12

TUESDAY — Number Sense

1. What is the greatest number you can make from these digits:

    3 7 4 1

2. Add: 87.4 + 34.8

3. Write 29 879 in words.

4. Write the following in standard form:

    20000 + 3000 + 100 + 9 + 0.4

5. Adam ate one quarter of his mom's apple pie. If it had twelve pieces, how many pieces did he eat?

Week 17

WEDNESDAY — Geometry

1. How many vertices does a triangular-based pyramid have?

2. Name a shape that is not a polygon.

3. Name two shapes with less than 5 sides.

4. Slide this shape to the right.

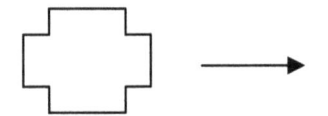

5. Classify the angle.

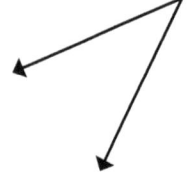

THURSDAY — Measurement

1. How many decades in 7 centuries?

1. How many seconds are in 21 minutes?

2. What unit of measurement would you use to find the temperature outside?

3. 301 dm = _____ mm

4. What is the perimeter and area of Mr. Levy's field? It measures 40 m wide and 12 m long.

FRIDAY — Data Management

Samantha's baseball team sold popsicles at all the league games to raise money for their team. The chart below shows how many popsicles they sold at each game.

Game	Popsicles Sold
Game 1	🍭🍭🍭
Game 2	🍭🍭🍭🍭🍭
Game 3	🍭🍭
Game 4	🍭🍭🍭🍭🍭
Game 5	🍭🍭🍭🍭

🍭 = 5 popsicles

1. How many popsicles did they sell all season? _____

2. During which game did they sell the most? _____

3. How many did they sell? _____

4. At which game did they sell the least? _____

5. What is the range of the data? _____

6. What is the mean amount of popsicles sold in the season? _____

BRAIN STRETCH

At the school bake sale, Class A sold 11 dozen cookies and Class B sold 17 dozen cookies. How many more cookies did Class B sell?

MONDAY — Patterning and Algebra

1. Complete the following:

    395, 399, 403, 407, ___, ___, ___

2. Write the first three numbers for this pattern rule:

    start at 177, subtract 11

3. What will be the 21st shape in this pattern?:

    ➡ ⬆ ⬇ ➡ ⬆

4. Each spider has 8 legs. How many legs do 17 spiders have?

5. $45 \div \_\_ = 9$

TUESDAY — Number Sense

1. Write four and eighteen hundredths in numbers.

2. Write 34 011 in expanded form.

3. Subtract 60.59 - 32.34 :

4. Put the following numbers in order from least to greatest:

    1, 0.8, 1.1, 1.01

5. Don bought three books at the book store. Two of them were $9.99 each and one was $12.99. What was the total for all three books?

WEDNESDAY — Geometry

1. Reflect this shape.

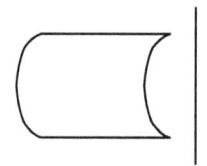

2. What kind of polygon is a square?

3. Look at the shapes. Choose flip, slide or turn.

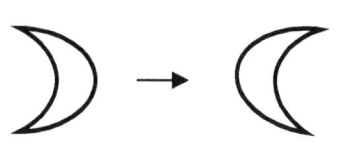

4. What is the measure of the missing angle?

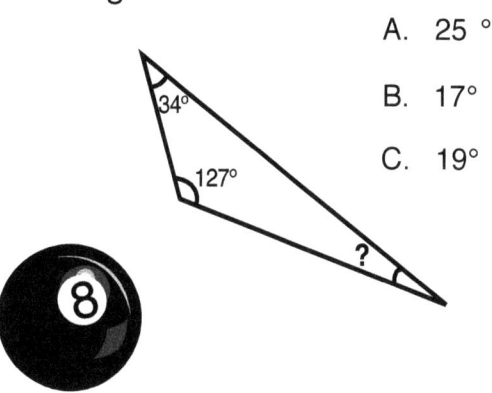

   A. 25°
   B. 17°
   C. 19°

   (triangle with angles 34°, 127°, ?)

5. What 3D figure does this object look like?

THURSDAY — Measurement

1. Draw a rectangle with a perimeter of 18 cm.

2. What device or tool would you use to find the size of an angle?

3. What time is it?

4. What unit of measurement would you use to find the mass of a car?

5. 25 litres = _____ millilitres

Week 18

FRIDAY — Data Management

Katherine took a survey of people's favourite gelato flavours. Here are the results.

Favourite Gelato Flavours

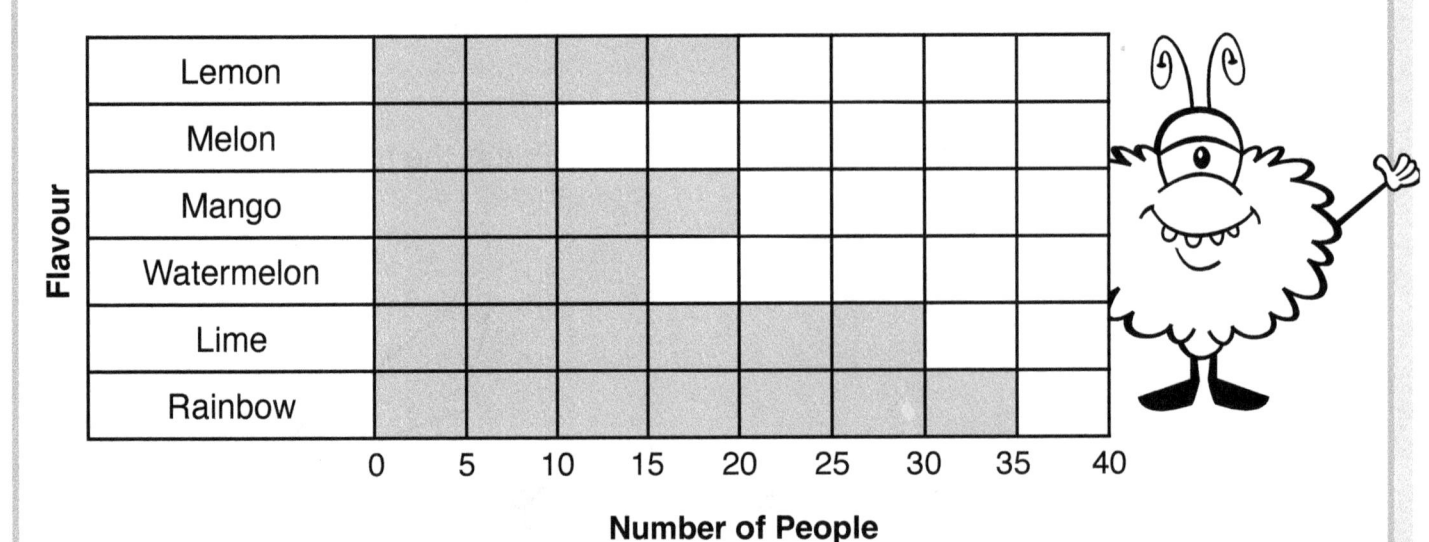

Flavour (y-axis):
- Lemon: 20
- Melon: 10
- Mango: 15
- Watermelon: 25
- Lime: 30
- Rainbow: 35

Number of People (x-axis: 0, 5, 10, 15, 20, 25, 30, 35, 40)

1. How many people were surveyed altogether? _____

2. What was the most popular gelato flavour? _____

3. What was the range of the data collected? _____

4. What was the mean of the data collected? _____

5. What was the least favourite gelato? _____

BRAIN STRETCH

The Morris family traveled the 341km to Watertown to visit their great aunt. They used 0.15L of fuel per kilometer.

a) How much fuel did they use altogether?

b) If the price of gas is 80 cents per L, how much money did they spend on fuel?

MONDAY — Patterning and Algebra

1. What is the rule for the following pattern?

   1, 4, 16, 64, 256, 1024

2. $64 \div a = 8$

   $a = $ _____

3. Complete the following:

   67, 71, 75, 79, ___ , ___ , ___

4. Is this a growing, shrinking or repeating pattern?

   316, 320, 324, 328, 332, 336

5. Write the first three numbers for this pattern rule:

   start at 424, subtract 12

TUESDAY — Number Sense

1. Which fraction is equivalent to $2/3$?

   $1/5$ $4/6$ $4/8$

2. Multiply: 34 x 91

3. Write <, > or = to make the expression true.

   4382 ☐ 4283

4. What is the value of the underlined digit?

   349<u>9</u>1

5. Kristine used a $10.00 bill to pay for her $2.78 pizza slice. How much change did she get back?

Week 19

WEDNESDAY — Geometry

1. An angle of 140° is called:

   A. obtuse   B. acute   C. right

2. How many edges does a cone have?

3. How many equal sides does an equilateral triangle have?

4. Which pair of shapes looks congruent?

   Q   R   S   T

   A. R and T   B. Q and S   C. Q and T

5. Which figure shows a line of symmetry?

   A.   B.   C.

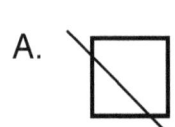

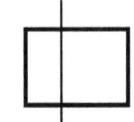

THURSDAY — Measurement

1. 2.3 m = _____ cm

2. What is the year 1 decade before 1968?

3. What would you use to measure the length of a soccer field?

4. What time is it?

5. Each side of a square measures 6 cm. What would the perimeter be if the sides were to increase by 5 cm each?

FRIDAY — Data Management

Use the calendar to answer the questions.

June

Sunday	Monday	Tuesday	Wednesday	Thursday	Friday	Saturday
				1	2	3
4	5	6	7	8	9	10
11	12	13	14	15	16	17
18	19	20	21	22	23	24
25	26	27	28	29	30	

1. How many Mondays are in the month of June? _____

2. What day of the week is June 24th? _____

3. Name the date that is 3 weeks after June 3rd _____

4. What is the date on the second Wednesday in June? _____

5. What day of the week will July start on? _____

BRAIN STRETCH

Matthew has a $36 monthly allowance. He puts away ½ of his money for savings and ¼ of his money on movie tickets. How much money does Matthew have left over?

Week 19

MONDAY — Patterning and Algebra

1. $56 \div b = 7$

    $b =$ _____

2. Write the first three numbers for this pattern rule:

    start at 91, add 7

3. What is the rule for the following pattern?

    7, 14, 21, 28, 35, 42

4. What will be the 19th number in this pattern?

    25, 30, 35, 40, 45, 50

5. Which pair of numbers best completes the equation?   ☐ X 100 = ☐

    A. 8.5 and 850   B. 8.5 and 8500   C. 85 and 850

TUESDAY — Number Sense

1. Write <, > or = to make the expression true.

    5541 ☐ 5514

2. Round this number to the nearest hundred.

    123 455

3. Orchard Park Elementary School has 89 primary students, 97 junior students and 67 intermediate students. How many students are at the school altogether?

4. Multiply: 11 x 82

5. Put the following numbers in order from least to greatest.

    6.6, 61/2, 61/4

WEDNESDAY — Geometry

1. What is a pentagon?

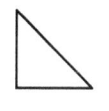

2. Reflect this shape.

3. Classify the angle.

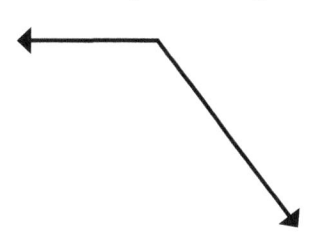

4. How many edges does a cylinder have?

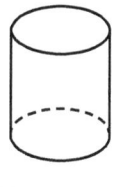

5. What 3D figure does this object look like?

THURSDAY — Measurement

1. How many seconds in 30 minutes?

2. 5.4 cm = _____ dm

3. What device or tool would you use to measure the width of a pencil?

4. How many hours in 12 days?

5. Choose the shape that would be the best unit to measure this area.

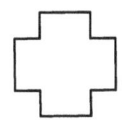

   A. ○   B. △   C. □

FRIDAY — Data Management

Each Friday, Mrs. Boffo draws a name from a prize jar and awards one of her students for their class participation. To get their name in the jar, students have to complete their work and participate in class activities. This Friday, the jar had 30 slips:
Liam - 5 slips, Kate - 15 slips, Mark - 3 slips, Julia - 7 slips

1. Who has the greatest probability of being chosen? _____

2. Who has the least probability of being chosen? _____

3. What is the probability of Liam being chosen? _____

4. What is the probability of Kate being chosen? _____

5. What is the probability of Julia being chosen? _____

BRAIN STRETCH

Lisa has to travel 763 km to visit her grandparents. She can drive on the highway at a speed of 100km/h or she can take the country roads and drive at 60 km/h.

a) How long will it take her to get to her grandparents' house if she takes the highway?

b) How long will it take her to get to her destination if she takes the country roads?

c) Why might she choose the slower option?

MONDAY — Patterning and Algebra

1. What kind of pattern is this?

    200, 170, 140, 110, 80

2. Complete the following:

    91, 85, 79, 73, ___, ___, ___

3. 8 x a = 88

    a = ___

4. Write the first three numbers for this pattern rule:

    start at 3, multiply by 5

5. What is the rule for the following pattern:

    50, 150, 250, 350, 450

TUESDAY — Number Sense

1. Write 70 878 in expanded form.

2. Add: $130.44 + $34.50

3. 2 cm of snow fell on average everyday for 4 weeks. How many mm fell after the 4 weeks?

4. What is the fractional part of the following number?

    ¼ of 24

5. Michael bought a computer game for $45.62. He paid the cashier $60.00. What was his change?

Week 21

WEDNESDAY — Geometry

1. Name two shapes with parallel lines.

2. Are these figures congruent or similar?

3. Draw a hexagon.

4. How many lines of symmetry does this letter have?

    P

5. Draw a 180 degree angle.

THURSDAY — Measurement

1. What temperature is it most likely to be if you are swimming in a lake during the summer?
    - A. 10°
    - B. 30°
    - C. 0°

2. What unit of measurement would you use to find the mass of a fly?

    A. kg   B. mg   C. g

3. 1100m=_____km

4. A round table has a diameter of 1.5 m. Which is the best estimate of the table's circumference?

    A. 4.7 m   B. 8 m   C. 3.5 m

5. The time is 11:15 p.m. What time will it be in 1 hour and 10 minutes?

FRIDAY — Data Management

The mayor of Moose Town holds a fishing derby each summer. The biggest catch was awarded the golden catfish trophy and a new rowboat. Here are the results of this year's derby.

Johnny	Robin	Leslie	Morgan	Melanie	Lauren	Tracy
2 kg	3.1 kg	2.8 kg	3.5 kg	3.2 kg	2.9 kg	1.2 kg

1. Put the masses in order from largest to smallest.
   _____

2. Who won the derby? _____

3. What is the range of the masses of fish? _____

4. What is the mean mass of the fish? _____

5. What is the median mass of the fish? _____

BRAIN STRETCH

Suzy can pedal 0.4 km in a minute.

a) How many km can she pedal in 10 minutes?

b) How many km can she pedal in 100 minutes?

c) How long will it take her to go 10 km?

d) How long will it take her to go 100km?

Week 21

MONDAY — Patterning and Algebra

1. What is the missing number?

    12 X _____ = 144

2. What is the rule for the following pattern?

    1, 3, 9, 27, 81, 243, 729

3. Write the first three numbers for this pattern rule:

    start at 1000, subtract 65

4. There were 420 people who attended a movie. If each ticket cost $6.00, how much money was collected?

5. What will be the 17th number in this pattern?

    1, 6, 11, 16, 21, 26, 31

TUESDAY — Number Sense

1. Put the following numbers in order from least to greatest:

    1.22, 1.5, 1.6, 1.1

2. Write two hundred eleven and ninety-nine hundreths in numbers.

3. Write <, > or = to make the expression true.

    12701 ☐ 12710

4. Luke rides his bike 7.4 km every week. How many kilometres does he ride in a year?

5. Divide: 678 by 3

WEDNESDAY — Geometry

1. Draw a right angle triangle.

2. What is a rhombus?

3. Draw a scalene triangle.

4. Draw an isosceles triangle.

5. Draw an obtuse angle.

THURSDAY — Measurement

1. What device or tool would you use to measure the speed of a race?

2. What time is it?

3. What unit of measurement would you use to find the mass of a vitamin?

4. 75 dm = _____ km

5. The time is 1:34 p.m. What time will it be in 4 hours and 10 minutes?

Week 22

FRIDAY — Data Management

Complete a line graph using the data from the table.

Soccer Game Attendance

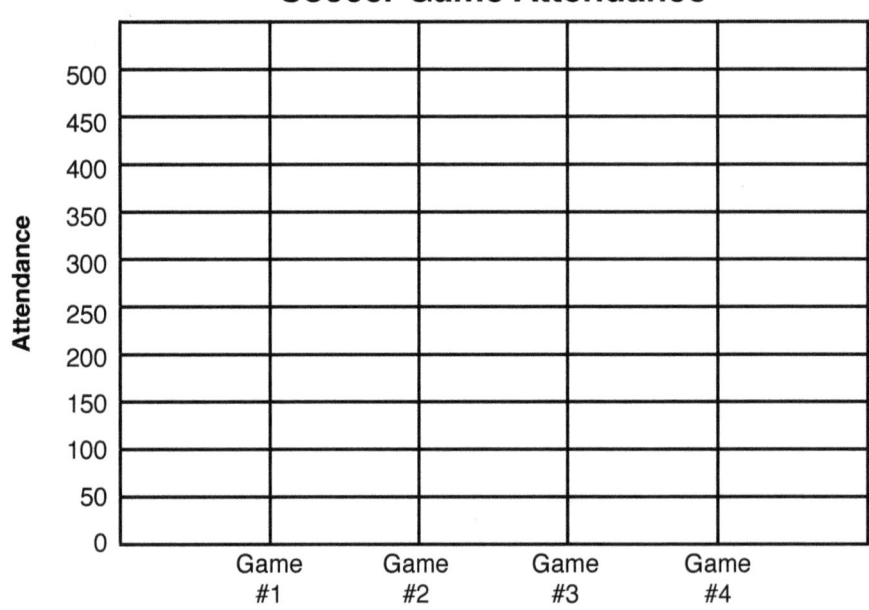

(Y-axis: Attendance, 0 to 500 in increments of 50; X-axis: Game #1, Game #2, Game #3, Game #4)

Game	# of People
Game # 1	150
Game # 2	300
Game # 3	450
Game #4	200

1. Which game had the most attendance? _____

2. Which game had the least attendance? _____

3. Which game(s) did not have an attendance of at least 350 people? _____

4. Which game(s) had an attendance of more than 400 people? _____

BRAIN STRETCH

a) $0.5 \div 100 =$

b) $0.9 \div 100 =$

c) $44.9 \div 10 =$

d) $240 \div 10 =$

e) $5 \div 1000 =$

f) $2.7 \div 100 =$

MONDAY — Patterning and Algebra

1. Which number sentence has the same product as 12 X 4?

   A. 5 X 9
   B. 6 X 8
   C. 7 X 6

2. Write the first three numbers for this pattern rule:

   start at 20, add 6

3. What is the rule for the following pattern?

   50, 51, 53, 56, 60, 65, 71

4. Is this a growing, shrinking or repeating pattern?

   900, 880, 860, 900, 880

5. $81 \div 9 =$ _____

TUESDAY — Number Sense

1. Which fraction is equivalent to $\frac{1}{4}$?

   $\frac{2}{5}$ $\frac{2}{6}$ $\frac{2}{8}$

2. Subtract

   $440.69 - $50.90

3. Round this number to the nearest ten.

   178 322

4. What is the value of the underlined digit?

   <u>3</u>7 903

5. Connor trained for the track meet by running 2.4 km every day. How many kilometers did Connor run in a week?

Week 23

WEDNESDAY — Geometry

1. Which 3D figure has 12 edges?

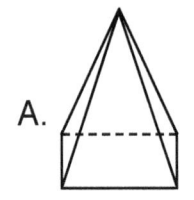

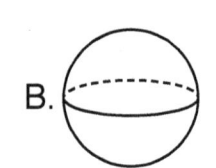

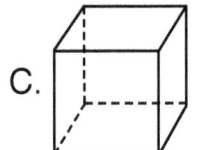

   A. (pyramid)   B. (sphere)   C. (cube)

2. How many lines of symmetry does this letter have?

   B

3. Draw a rhombus.

4. Draw 2 intersecting lines.

5. Which figure shows a line of symmetry?

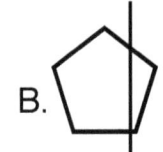

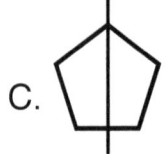

   A.   B.   C.

THURSDAY — Measurement

1. Find the volume of this box.

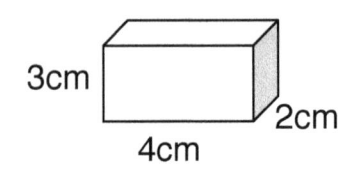

   3cm, 2cm, 4cm

2. How many seconds are in a quarter hour?

3. Each side of a square measures 9 cm. What would the perimeter be if the sides were to decrease by 3 cm?

4. What unit of measurement would you use to find your height?

5. 10.33 cm = _____ dm

Week 23

FRIDAY — Data Management

Mrs. Gray has a jar full of marbles on her desk. The jar has 20 blue marbles, 50 white marbles, 30 yellow marbles, 40 red marbles. 40 pink marbles and 20 green marbles.

1. What colour marble is most likely to be randomly picked from the jar?
   _____

2. What two colours of marbles are least likely to be randomly picked from the jar?
   _____

3. What is the probability of selecting a yellow marble? _____

4. What is the probability of selecting a blue marble? _____

5. Is it more likely to select a pink or a yellow marble? _____

BRAIN STRETCH

The Dixie Mart Snack Shop sells pizza slices for $2.10, pop for $1.50 and chips for $0.75.

a) How much would it cost to buy a family of four a pizza slice, pop and chips for each person?

b) How much change would they get back from two ten dollar bills?

c) What would be the fewest number of bills or coins needed to make the change?

Week 23

MONDAY — Patterning and Algebra

1. 14 x a = 42

   a = _____

2. What is the rule for the following pattern?

   1000, 900, 850, 750, 700, 600

3. What will be the 14th shape in this pattern?

   ♡ ◇ ☺ ♡

4. Write the first three numbers for this pattern rule:

   start at 2 multiply by 3 and subtract 1

5. Complete the pattern:

   505, 500, 490, 470, ___, ___, ___

TUESDAY — Number Sense

1. Write seven hundred fifteen dollars and sixteen cents in numbers.

2. Write 606.55 in expanded form.

3. Write an equivalent fraction to $\frac{4}{5}$.

4. What is the value of the underlined digit?

   3<u>8</u>4 500

5. Jack has 284 hockey cards. He promised to give ¼ of them to his sister. How many cards will he give her?

WEDNESDAY — Geometry

1. Which pair of shapes look congruent?

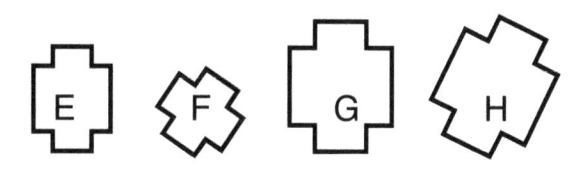

   E   F   G   H

   A. F and H   B. G and H   C. E and F

2. How many veritices does a circle have?

3. Classify the following pair of lines.

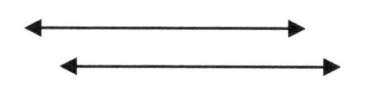

   A. intersecting   B. parallel   C. perpendicular

4. Classify the angle.

5. Draw a cube.

THURSDAY — Measurement

1. 90 km = _____ cm

2. It takes Lewis 35 minutes and 40 seconds to walk to the park. How long does a round trip take?

3. Find the volume of a box measuring 10mm x 10mm x 3mm.

4. How many decades in 4 centuries?

5. What unit of measurement would you use to find the space between your eyes?

Week 24

FRIDAY — Data Management

Nina made extra money in the summer by washing dogs for neighbourhood clients. She kept track of the number of dogs she washed in the chart below.

Week	Number of Dogs
1	🐕 🐕 🐕
2	🐕 🐕
3	🐕 🐕 🐕 🐕
4	🐕
5	🐕 🐕

🐕 = 5 dogs

1. How many dogs did Nina wash altogether? _____

2. If each dog wash costs $8 how much money did Nina make? _____

3. What is the range of the data? _____

4. What is the mean amount of dogs that she washed each week? _____

BRAIN STRETCH

a) Megan bought 6 T-shirts. Each T-shirt cost $6.60. How much did the T-shirts cost altogether?

b) She paid with 2 twenty dollar bills. Did she get change? Explain.

MONDAY — Patterning and Algebra

1. What is the rule for the following pattern?

   2, 20, 200, 2000, 20000

2. $125 - a = 58$

   $a =$ _____

3. Complete the pattern:

   17, 18, 20, 23, __, __, __

4. What will be the 10th number in this pattern?

   79, 77, 75, 73, 71

5. Is this a growing, shrinking or repeating pattern?

   112, 124, 136, 148

TUESDAY — Number Sense

1. Write 1011 in words.

2. Write <, > or = to make the expression true.

   3197 ☐ 3791

3. Multiply: 167 x 30

4. Trisha wants to bring cookies to share with her 27 classmates. If a package has 12 cookies, how many packs does he need to bring?

5. Write the following in standard form:

   $3000 + 100 + 4 + 0.9 + 0.07$

Week 25

WEDNESDAY — Geometry

1. Classify this angle.

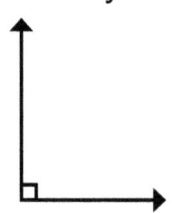

2. Draw an acute triangle.

3. Rotate this shape 90 degrees.

4. Measure this angle.

5. How many sides does a decagon have?

THURSDAY — Measurement

1. Mary's paper route takes her about 1 hour and 10 minutes each day. How long does she spend delivering papers in a week?

2. What unit of measurement would you use to find the duration of a soccer game?

3. 64 mm = _____ cm

4. How many weeks are there in four years?

5. How many days in 5 years?

FRIDAY — Data Management

Use the spinner below to answer the questions that follow.

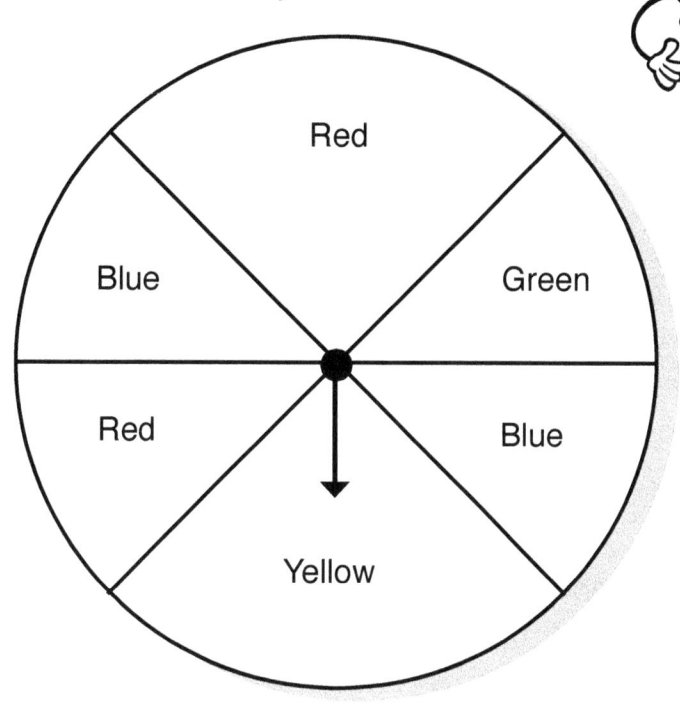

Spinner sections: Red, Green, Blue, Yellow, Red, Blue

1. What colour are you most likely to land on? _____

2. What colour are you least likely to land on? _____

3. What is the probability that you will land on green? _____

4. What is the probability that you will land on yellow? _____

5. What is the probability that you will land on blue? _____

BRAIN STRETCH

Brian rides 2.6 km to and 2.6 km from school every day. How many kilometers will he ride in a week?

Week 25

MONDAY — Patterning and Algebra

1. In which number sentence does a 9 make the equation true?

    A. 5 x _____ = 40

    B. 21 ÷ _____ = 3

    C. 24 + _____ = 33

2. Create a repeating pattern.

3. $79 - b = 22$

    $b =$ ____

4. What is the rule for the following pattern?

    5, 25, 125, 625, 3125

5. Write the first three numbers for this pattern rule:

    start at 4, x 2 -1

TUESDAY — Number Sense

1. Tyler can swim one length of his pool in 25 seconds. How many lengths can he swim in five minutes?

2. Write an equivalent fraction to $\frac{1}{10}$.

3. Divide 4.5 by 100:

4. Put the following numbers in order from greatest to least

    200, 220, 22, 202

5. What is the value?

WEDNESDAY — Geometry

1. Name a shape with more than 6 sides.

2. Draw an acute angle.

3. What shape is a cone's face?

4. Draw an obtuse angle.

5. Complete:

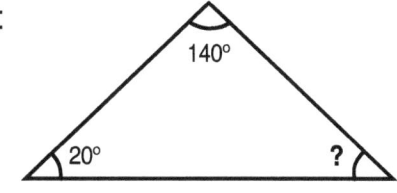

   (triangle with angles 140°, 20°, and ?)

   measure of angle _____

   type of triangle _____

THURSDAY — Measurement

1. The time is 6:14 a.m. What time will it be in 76 minutes?

2. What device or tool would you use to measure the volume of some milk?

3. What unit of measurement would you use to find the capacity of a can of soda pop?

4. How much time has elapsed between:

   12:11:30 and 15:24:55?

5. 780 dm = _____ km

Week 26

FRIDAY — Data Management

Deborah and her cousins raced remote control cars. They created a course and used a timer to measure the time it took each car to finish the track. Here are their results.

Deborah	Mary	Lou	Leonard	Michael
76 secs.	98 secs.	81 secs.	65 secs.	70 secs.

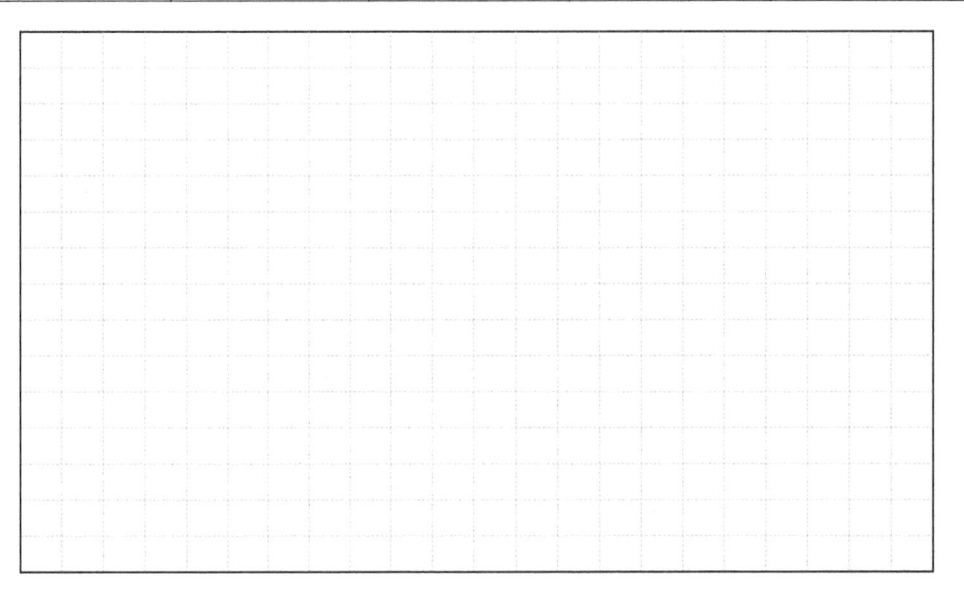

1. Create a bar graph in the space provided to display the data collected.

2. Who had the best time? _____

3. Who had the slowest time? _____

4. What is the range of the data? _____

5. What is the mean time? _____

BRAIN STRETCH

Write the factors for each given number.

a) 72

b) 15

MONDAY — Patterning and Algebra

1. Write the first three numbers for this pattern rule:

   start at 65, subtract 3

2. What will be the 11th number in this pattern?

   12, 14, 16, 18, 20, 22

3. What is the rule for the following pattern?

   90000, 9000, 900, 90, 9, 0.9

4. Complete the pattern:

   22, 44, 66, 88, ___, ___, ___

5. Which number sentence has the same quotient as 56 ÷ 8

   A. 32 ÷ 4   B. 49 ÷ 7   C. 27 ÷ 3

TUESDAY — Number Sense

1. List the factors for 35:

2. Write <, > or = to make the expression true.

   5005 ☐ 5055

3. Write 89 266 in expanded form.

4. Add 455 + 2390

5. A box of books can hold 55 books. How many boxes are needed for 1140 books?

Week 27

WEDNESDAY — Geometry

1. Draw a triangle with a 60 degree interior angle.

2. How many edges does a cylinder have?

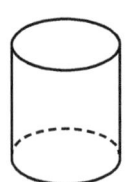

3. Slide this shape 3 cm to the right.

4. What kind of triangle has sides of different length?

5. What is a transformation?

THURSDAY — Measurement

1. What would the side length be for a square with a perimeter of 100 cm.

2. 900 mm = _____ km

3. How many rectangles can you draw with an area of 36 cm^2?

4. What unit of measurement would you use to find the capacity of your kitchen sink?

5. How many hours are there in 6 days?

FRIDAY — Data Management

Answer the following probability questions with: very likely, likely, or not likely.

1. What is the likelihood that you will travel by airplane to school tomorrow?

   _____

2. What is the likelihood that you will read a book this week?

   _____

3. What is the likelikhood that you will eat lunch tomorrow?

   _____

4. What is the likelihood that you will meet a talking bear?

   _____

5. What is the likelihood that you will answer a math question today?

   _____

BRAIN STRETCH

Brandon's dad gave him $60 to buy a new baseball glove and running shoes.

a) If he chooses to buy a $43 baseball glove, how much money would he have left to spend on shoes?

b) If he chooses to buy $37 dollar running shoes, how much money would he have left to buy a baseball glove?

c) Give an example of what denomination of bills or coins Brandon's dad could have used to give him the money.

d) How much money is half of $60?

Week 27

MONDAY — Patterning and Algebra

1. 64 − 29 = a

   a = _____

2. What is the rule for the following pattern?

   44, 45, 47, 50, 54, 59, 65

3. Complete the following pattern:

   98, 76, 98, 76, ___, ___, ___

4. Write the first three numbers for this pattern rule:

   start at 704, add 20

5. Vicki walks 4km every day. How many km will Vicki walk after 21 days?

TUESDAY — Number Sense

1. Write the value of the underlined digit.

   <u>8</u>7 601

2. Write fifty thousand two hundred nineteen and two tenths in numbers.

3. Madeline spent $72.35 at the grocery store. How much change did she get back from four twenty dollar bills?

4. Write an equivalent fraction to ½.

5. Round this number to the nearest tenth.

   5678.46

WEDNESDAY — Geometry

1. An angle of 160° is called:

2. How many edges does a rectangular prism have?

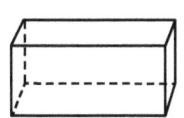

3. Draw a triangle with a line of symmetry.

4. Draw a shape with a line of symmetry.

5. How are a square and rhombus alike?

THURSDAY — Measurement

1. Serge plays hockey for 1 hour and 20 minutes each week. How much hockey does he play in a year?

2. What device would you use to find your mass?

3. 2.5 m = _____ km

4. What is the perimeter of a square with 12 m sides?

5. What unit of measurement would you use to find the length of time it takes to get to your school?

FRIDAY — Data Management

Oakridge High School sold magazine subscriptions to raise money for school trips. This graph shows the number of subscriptions sold by each class.

Magazine Sales

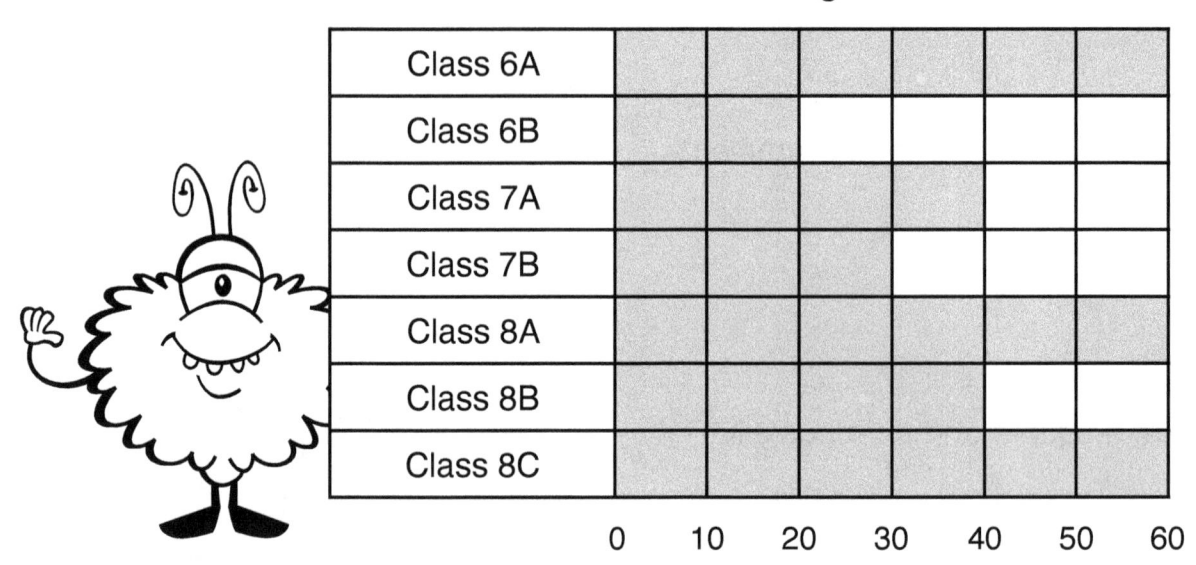

Class	0	10	20	30	40	50	60
Class 6A							
Class 6B							
Class 7A							
Class 7B							
Class 8A							
Class 8B							
Class 8C							

Number of Magazine Subscriptions

1. Which classes sold the most subscriptions? _____

2. Which class sold the least number of subscriptions? _____

3. What is the range of the data? _____

4. What is the mean number of subscriptions sold? _____

5. What is the median? _____

BRAIN STRETCH

How much money in total?

12 twenty-dollar bills, 1 toonie, 3 loonies, 1 quarter, 2 dimes, and 5 nickels.

MONDAY — Patterning and Algebra

1. Write the first three numbers for this pattern rule:

   start at 100, add 14

2. $276 - b = 198$

   b = _____

3. What is the rule for the following pattern?

   701, 711, 721, 731, 741, 751

4. What will be the 15th shape in this pattern?

   □ ○ ▽ □

5. Complete the following pattern:

   12, 121, 1212, 12121, _____, _____, _____

TUESDAY — Number Sense

1. Write <, > or = to make the expression true.

   334 □ 334

2. Write 22 190 in words.

2. Subtract 24 from 303:

3. Tenzin delivers 80 flyers per week. How many does she deliver in a year?

5. Write an equivalent fraction to 1/3.

Week 29

WEDNESDAY — Geometry

1. Reflect this shape.

2. How many edges does a square-based pyramid have?

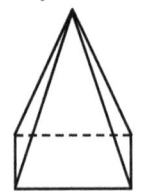

3. What is an equilateral triangle?

4. What 3D shape can be made from this net?

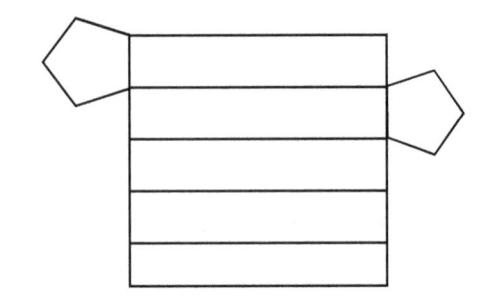

5. Draw a triangle with a 20 degree angle.

THURSDAY — Measurement

1. What tool would you use to measure the temperature outside?

2. What unit of measurement would you use to find the duration of one school year?

3. 49 dm = _____ km

4. Draw a triangle with an area of 8mm^2

5. How much time has elapsed between 14:37:30 and 20:50:00?

FRIDAY — Data Management

Jonathon has a paper route and delivers the papers every weekday during the year. At the end of each month, he has to collect money for the papers he has delivered in order to pay his supplier. Here is a list of how much money he collected during the first six months of the year.

Month	Amount Collected
January	$310
February	$278
March	$305
April	$305
May	$298
June	$314

1. What is the range of the money he collected? _____

2. What is the mean amount collected? _____

3. What is the median amount collected? _____

4. What is the most he collected? _____

5. Why would he have collected the least in February?
   _____

BRAIN STRETCH

a) How many donuts do you get when you buy 25 dozen?

b) How much did each donut cost if you paid $23.00?

Week 29

MONDAY — Patterning and Algebra

1. What is the rule for the following pattern?

    499, 598, 697, 796, 895

2. Is this a growing, shrinking or repeating pattern?

    76, 71, 66, 61, 56, 51

3. $500 \div a = 10$

    a=___

4. Complete the following pattern:

    701, 652, 603, 554, ___, ___

5. Write the first three numbers for this pattern rule:

    start at 1000, divide by 2

TUESDAY — Number Sense

1. Multiply: 23 x 55

2. Divide: 2.89 by 100

3. George has a paper airplane that can fly 3.6 m. His friend Paul has one that can go 246 cm. How much further can George's plane fly?

4. Put the following numbers in order from least to greatest.

    7.7, 1.7, 7.2, 7.1

5. Write 99 807 in expanded form.

WEDNESDAY — Geometry

1. What kind of translation is seen here?

   ⇒ | ⇐

2. Draw a triangle with a 100 degree angle.

3. How many edges does a triangular-based pyramid have?

4. Draw a set of parallel lines.

5. How are a rhombus and hexagon different?

THURSDAY — Measurement

1. Alexander practices her violin for 45 minutes each day. How long does he practice each week?

2. How many years are in one century?

3. What unit of measurement would you use to find the mass of a feather?

4. 6.7 kg = _____ mg

5. What is the year 5 decades before 1969?

FRIDAY — Data Management

1. How do you find the mean average of a set of data? _____
   _____

2. How do you find the median of a set of data? _____
   _____

3. What is the range of a set of data? _____
   _____

4. What is a tally chart? _____
   _____

5. What parts of a bar graph should be labeled? _____
   _____

BRAIN STRETCH

Bernie's part time job at a community centre included filling a soda machine with cans of pop.

a) If the machine can hold 300 cans of pop, approximately how many cases of 24 cans of pop must Bernie purchase to fill the machine?

b) If each can is to be sold for $1.25, how much money should Bernie collect when the machine is empty?

c) What is the fewest number of coins needed to buy a soda?

Math — Show What You Know!

_____

- [ ] I read the question and I know what I need to find.
- [ ] I drew a picture or a diagram to help solve the question.
- [ ] I showed all the steps in solving the question.
- [ ] I used math language to explain my thinking.

Student Tracking Sheet

Student	Week 1	Week 2	Week 3	Week 4	Week 5	Week 6	Week 7	Week 8	Week 9	Week 10	Week 11	Week 12	Week 13	Week 14	Week 15

Student Tracking Sheet

Student	Week 16	Week 17	Week 18	Week 19	Week 20	Week 21	Week 22	Week 23	Week 24	Week 25	Week 26	Week 27	Week 28	Week 29	Week 30

© Chalkboard Publishing

Student Certificate

You Are Incredible!

Keep Up the Good Work!

_____

Week 1

Mon.   1. 68   2. 3   3. 12, 17, 23   4. start at 5, add 3   5. 90, 86, 82, 78…
Tues.   1. nine thousand five hundred forty-five   2. 0.01, 0.11, 1.01, 1.10
3. 30 000+4000+900+80+7   4. 6.9   5. $361.15
Wed.   1. b   2. none   3. hexagon   4. 4   5. 6
Thurs.   1. 6000 mm   2. 3:00 p.m.   3. 355 ml   4. kg   5. 20cm
Fri.   1. 20   2. car   3. 8   4. 4   5. see graph paper
Brain Stretch   a. 280   b. 42

Week 2

Mon.   1. 48, 60, 72   2. start at 88, subtract 11   3. 8   4. circle   5. 2, 15, 28, 41
Tues.   1. 2.02, 2.11, 21   2. 12 007   3. 350   4. $10.12   5. 30
Wed.   1. rectangle or parallelogram   2. an angle greater than 90°and less than 180 °   3. 6
   4. 3   5. cylinder
Thurs.   1. 14.6 cm   2. 20 dm   3. cm   4. 7:30   5. 11:59 a.m.
Fri.   1. move forward   2. 1/2   3. 3/8   4. 1/8
Brain Stretch   190 km

Week 3

Mon.   1. 8   2. 67, 63, 59, 55   3. 71, 61, 51   4. start at 12, add 2   5. repeating
Tues.   1. 3000+400+30+2   2. two thousand four hundred fifty   3. 11   4. 8431   5. $5.34
Wed.   1. 90° angle   2. 5   3. circle   4. b- cone   5. 5
Thurs.   1. 2:25 p.m.   2. 55 m^2 3. 1.2 cm   4. mg   5. a
Fri.   1. see graph   2. 65   3. skipping ropes   4. hula hoops   5. 40
Brain Stretch   a. $1560 per year   b. $780

Week 4

Mon.   1. 900, 915, 930, 945   2. 20, 26, 33   3. start at 200, add 25   4. 33   5. 15
Tues.   1. 27   2. 990, 909, 0.99, 0.09   3. 55 000   4. 42.12   5. b
Wed.   1. 1   2. 4- 90° angles   3. less than 90 °   4. cube   5. parallelogram
Thurs.   1. metres   2. 2 500 000 m   3. 11:10   4. 110 cm   5. 10 cm
Fri.   1. 63 marbles   2. 17/63   3. 3/63 or 1/21   4. yellow   5. clear
Brain Stretch   a. 500   b. 520   c. 225   d. 2400

Week 5

Mon.   1. start at 70, subtract 5   2. 129   3. 4, 29, 54, 79   4. 96, 192, 384   5. sun
Tues.   1. 263.7   2. 91   3. two thousand fifty-two   4. 7209.6   5. $41.98
Wed.   1. 2   2. trapezoid   3. 2   4. ←—•—→   5. rectangular prism
Thurs.   1. 7:45 p.m.   2. answers will vary i.e.- 3 cm, 3 cm,3cm,   3. 180 m^2 4. 56 m   5. 8000mm
Fri.   1. Bert   2. Maria   3. 14   4. 6   5. 9.6
Brain Stretch   a. $1.50   b. 5

Week 6

Mon. **1.** growing **2.** 222, 204, 186, 168 **3.** 6 **4.** start at 45, add 3 **5.** 225, 275, 325
Tues. **1.** 4 **2.** 10 + 2 + 0.7 + 0.08 **3.** 0.22, 2.2, 20.2 **4.** 1850 **5.** 13
Wed. **1.** ans. will vary **2.** between 180 and 360 degrees **3.** square based pyramid **4.** circle **5.** 1
Thurs. **1.** 3300 mm **2.** b **3.** 540 minutes **4.** 16 cm2 **5.** 9:15
Fri. **1.** French fries **2.** salads **3.** 23 **4.** 13.5 **5.** 12.66 **6.** 76
Brain Stretch **a.** $15.10 **b.** $4.90

Week 7

Mon. **1.** 26, 29, 32 **2.** 27 **3.** 75 **4.** 7, 14, 28, 56 **5.** alternately add 11, then subtract 11
Tues. **1.** $100.68 **2.** 2358 **3.** 192 **4.** one thousand four hundred ninety-nine **5.** 3990
Wed. **1.** rectangular prism **2.** none **3.** diamond, rhombus, or parallelogram **4.** square **5.** ✕
Thurs. **1.** 5 m **2.** 0.066 m **3.** km **4.** c **5.** p = 20 units, a = 15 units^2
Fri. **2.** 183 **3.** soccer **4.** chess **5.** 36.6 **f.** 47
Brain Stretch **a.** hundreds **b.** tens **c.** hundreds **d.** ones **e.** thousands **f.** ten thousands

Week 8

Mon. **1.** 245, 256, 267, 278 **2.** start at 100, add 100 **3.** 25 **4.** shrinking **5.** 124, 135, 146
Tues. **1.** b **2.** 6790 **3.** 1260 L **4.** 77.7, 7.07, 7.0, 0.77 **5.** 184
Wed. **1.** square based pyramid **2.** square based pyramid **3.** c **4.** slide **5.** a. right
Thurs. **1.** the distance around a figure or object **2.** 12:20 **3.** 10 000 dm **4.** mg **5.** 595 cm
Fri. **1.** see diagram **2.** blue **3.** green and yellow **4.** green and yellow
Brain Stretch **a.** 7 bags **b.** $13.93

Week 9

Mon. **1.** start at 20, add 4 **2.** 20, 33, 46, 59 **3.** 199, 190 **4.** triangle **5.** 144
Tues. **1.** 1340 **2.** 8700 **3.** 600+ 90+ 8+ 0.2 **4.** 32 **5.** two thousand two hundred fifty
Wed. **1.** 8 **2.** hexagon **3.** none **4.** ▱ **5.** n and o
Thurs. **1.** cm **2.** l x w **3.** mins. **4.** 9.9 dm **5.** 1:49 a.m.
Fri. **1.** Montreal **2.** Victoria, Winnipeg or Halifax, North Bay **3.** 140 **4.** 80 **5.** 97.14
Brain Stretch 520, 5200, 52 000

Week 10

Mon. **1.** ans. will vary **2.** 39, 45, 51 **3.** start at 130, subtract 7 **4.** 100, 93, 86, 79 **5.** repeating
Tues. **1.** 546 **2.** 649.2 **3.** 9 **4.** 25 **5.** 0.55, 5.5, 5.55, 55
Wed. **1.** see drawing **2.** 12 **3.** 1 **4.** sphere **5.** 3 sets
Thurs. **1.** a = 8 m^2 p= 12 m **2.** b **3.** 32 **4.** 11:40 **5.** 0.5 m
Fri. **1.** 10/50 or 1/5 **2.** 6/50 or 3/25 **3.** stickers **4.** jacks or stuffed animals **5.** ans. will vary
Brain Stretch **a.** $0.16 **b.** 90 **c.** $14.40 **d.** $5.60

Week 11

Mon. **1.** 2, 4, 8, 16 **2.** 24 **3.** 41 **4.** 54, 67, 80 **5.** start at 400, add 10
Tues. **1.** nine thousand nine hundred one **2.** 80 930.8 **3.** 114 **4.** = **5.** $8.22
Wed. **1.** cylinder **2.** none **3.** ans. will vary **4.** 60 degree, equilateral **5.** b
Thurs. **1.** 30 000 **2.** 7:50 **3.** grams **4.** ans. will vary **5.** 0.8 kg
Fri. **1.** soap opera = 12, sports = 10, comedy = 16, drama = 6, cartoon = 20.5, news = 3 **2.** 22.5 **3.** cartoon **4.** 35 **5.** news
Brain Stretch **a.** 245 **b.** 4410 **c.** 12 740

Week 12

Mon. **1.** 256, 1024, 4096 **2.** start at 700, subtract 11 **3.** 9 **4.** shrinking **5.** 75, 90, 105, 120
Tues. **1.** 3.32, 3.22, 2.32, 2.23 **2.** 675 **3.** 15 **4.** 7000 + 600 + 40 + 4 + 0.3 **5.** 45
Wed. **1.** none **2.** 4 **3.** octagon **4.** b **5.** ans. will vary
Thurs. **1.** 9.66 m^2 **2.** 22.5 mm **3.** 0.0044 **4.** mm **5.** 720 min
Fri. **1.** strawberries **2.** bananas **3.** apples and grapes
Brain Stretch $44.65

Week 13

Mon. **1.** repeating **2.** 61, 69, 77, 85 **3.** 260, 200, 140 **4.** c **5.** start at 309, subtract 3
Tues. **1.** 23.2 **2.** seventy-eight thousand two hundred fifteen **3.** 4/8 **4.** 225 **5.** 61
Wed. **1.** cube **2.** 4 **3.** no **4.** a **5.** rectangular prism, 6 faces, 12 edges, 8 vertices
Thurs. **1.** 14.2 km **2.** 4 **3.** 6:15 **4.** 1.63 m **5.** 24 cm
Fri. **1.** 60 **2.** 4 **3.** 16 **4.** blue **5.** yellow
Brain Stretch **a.** $6.30 **b.** $13.70 **c.** 7

Week 14

Mon. **1.** 97, 104, 111 **2.** start at 557, add 2 **3.** 444 **4.** 89, 109, 129, 149 **5.** 278
Tues. **1.** 340 **2.** ¼, 1/3, 1/2 **3.** 60 043 **4.** 200 **5.** 79.89
Wed. **1.** b **2.** a **3.** 1 **4.** ans. will vary **5.** acute
Thurs. **1.** litres **2.** 90 000 m **3.** 2710 cm **4.** 8:55 **5.** 1600 cm^2
Fri. **1.** week 1 **2.** $22.00 **3.** $114.00 **4.** $28.00
Brain Stretch **a.** 1000 **b.** 10 000 **c.** 100 000

Week 15

Mon. **1.** 1000, 950, 900, 850 **2.** repeating **3.** 66, 85, 106 **4.** b **5.** start at 999, add 3
Tues. **1.** 0.79 **2.** 3182 **3.** 2193 **4.** 1000 + 9 + 0.8 + 0.08 **5.** 2/8
Wed. **1.** turn **2.** i.e. square, rhombus, rectangle, trapezoid **3.** 2 **4.** b **5.** 20°, isosceles
Thurs. **1.** 350 000 m **2.** 60 months **3.** $92.00 **4.** 9500 m **5.** 11:40
Fri. **1.** 32 **2.** 7 **3.** 3 **4.** 22 **5.** lands on side **6.** 7/32 **7.** 3/32
Brain Stretch **a.** 58 **b.** 4050 **c.** 21 635 **d.** 22 090 **e.** 5.2 **f.** 79

Week 16

Mon. **1.** penguin **2.** 79, 74 **3.** 600, 617, 634, 651 **4.** start at 246, subtract 5 **5.** 2
Tues. **1.** 31.9 **2.** 1450 **3.** 1 ¾, 1 ½, 1 ¼ **4.** 6900.7 **5.** 60
Wed. **1.** isosceles **2.** ⌐ **3.** ∟ **4.** triangle **5.** square based pyramid, 5, 8, 5
Thurs. **1.** 6:55 **2.** ans. will vary **3.** 12.5 cm **4.** cm **5.** measure the outside walls with a metre stick
Fri. **1.** 18 **2.** 21 **3.** 22.28 **4.** 8 **5.** 16
Brain Stretch **a.** 20 people **b.** 2 tubs

Week 17

Mon. **1.** 13 **2.** start at 50, multiply by 2 **3.** shrinking **4.** 56, 45 **5.** a
Tues. **1.** 7431 **2.** 122.2 **3.** twenty nine thousand eight hundred seventy-nine **4.** 23109.4 **5.** 3
Wed. **1.** 4 **2.** ans. will vary **3.** i.e. rectangle, triangle, rhombus **4.** ✛ **5.** acute
Thurs. **1.** 70 **2.** 1260 min **3.** celsius **4.** 30 100 **5.** p = 104 m, a = 480 m²
Fri. **1.** 100 **2.** Game 4 **3.** 30 **4.** Game 3 **5.** 20 **6.** 20
Brain Stretch 6 dozen or 72 cookies

Week 18

Mon. **1.** 411, 415, 419 **2.** 177, 166, 155, 144 **3.** ⌂ **4.** 136 **5.** 5
Tues. **1.** 4.18 **2.** 30 000 + 4000 + 10 + 1 **3.** 28.25 **4.** 0.8, 1, 1.01, 1.1 **5.** $32.97
Wed. **1.** ⌒ **2.** regular polygon **3.** flip **4.** c **5.** sphere
Thurs. **1.** ans. will vary **2.** protractor **3.** 2:25 **4.** kg **5.** 25 000 ml
Fri. **1.** 130 **2.** Rainbow **3.** 25 **4.** 21.66 or 21.7 **5.** melon
Brain Stretch **a.** 51.15 litres **b.** $40.92

Week 19

Mon. **1.** start at 1, multiply by 4 **2.** 8 **3.** 83, 87, 91 **4.** growing **5.** 424, 412, 400, 388
Tues. **1.** 4/6 **2.** 3094 **3.** > **4.** 90 **5.** $7.22
Wed. **1.** a **2.** none **3.** 3 **4.** b **5.** none
Thurs. **1.** 230 cm **2.** 1958 **3.** metre stick **4.** 2:45 **5.** p = 44 cm
Fri. **1.** 4 **2.** Saturday **3.** Saturday June 24th **4.** June 14th **5.** Saturday
Brain Stretch $9.00 left over

Week 20

Mon. **1.** 8 **2.** 91, 98, 105, 112 **3.** start at 7, add 7 **4.** 115 **5.** a
Tues. **1.** > **2.** 123 500 **3.** 253 **4.** 902 **5.** 6 ¼, 6 ½, 6.6
Wed. **1.** a 5 sided polygon **2.** △ **3.** obtuse or reflex angle **4.** none **5.** cone
Thurs. **1.** 1800 s **2.** 0.54 dm **3.** mm ruler **4.** 288 hours **5.** c
Fri. **1.** Kate **2.** Mark **3.** 5/30 or 1/6 **4.** 15/30 or 1/2 **5.** 7/30
Brain Stretch **a.** 7.63 hrs **b.** 12.72 hrs **c.** ans. will vary – safer, more scenic

Week 21

Mon. **1.** shrinking **2.** 67, 61, 55 **3.** 11 **4.** 3, 15, 75, 375 **5.** start at 50, add 100
Tues. **1.** 70, 000 + 800 + 70 + 8 **2.** $164.94 **3.** 560 mm **4.** 6 **5.** $14.38
Wed. **1.** ans. will vary **2.** congruent **3.** (hexagon) **4.** none **5.** (line with point)
Thurs. **1.** b **2.** b **3.** 1.1 km **4.** a **5.** 12:25 a.m.
Fri. **1.** 3.5, 3.2, 3.1, 2.9, 2.8, 2, 1.2 **2.** Morgan **3.** 2.3 **4.** 2.67 **5.** 2.9
Brain Stretch **a.** 4 km **b.** 40 km **c.** 25 min **d.** 250 min

Week 22

Mon. **1.** 12 **2.** start at 1, multiply by 3 **3.** 1000, 935, 870, 805 **4.** $2520.00 **5.** 81
Tues. **1.** 1.1, 1.22, 1.5, 1.6 **2.** 211.99 **3.** < **4.** 384.8 km **5.** 226
Wed. **1.** (triangle) **2.** (rhombus) **3.** a triangle without any equal sides **4.** a triangle with 2 equal sides **5.** an angle that is more than 90° and less than 180°
Thurs. **1.** stopwatch **2.** 3:40 **3.** mg **4.** 0.0075 **5.** 5:44 p.m.
Fri. **1.** game 3 **2.** game 1 **3.** games 1, 2 and 4 **4.** game 3
Brain Stretch **a.** 0.005 **b.** 0.009 **c.** 4.49 **d.** 24 **e.** 0.005 **f.** 0.027

Week 23

Mon. **1.** b **2.** 20, 26, 32, 38 **3.** start at 50, add 1, add 2, add 3… **4.** repeating **5.** 9
Tues. **1.** 2/8 **2.** $389.79 **3.** 178 320 **4.** 30 000 **5.** 16.8 km
Wed. **1.** c **2.** 1 **3.** see picture **4.** ans. will vary **5.** c
Thurs. **1.** 24 cm^3 **2.** 900 s **3.** 24 cm **4.** cm or m **5.** 1.033 dm
Fri. **1.** white **2.** blue or green **3.** 30/200 or 3/20 **4.** 20/200 or 1/10 **5.** pink
Brain Stretch **a.** $17.40 **b.** $2.60 **c.** 4 coins – 1 toonie, 2 quarters, 1 dime

Week 24

Mon. **1.** 3 **2.** start at 1000, subtract 100, then subtract 50, alternately **3.** diamond **4.** 2, 5, 14, 41 **5.** 430, 350, 190
Tues. **1.** $715.16 **2.** 600 + 6 + 0.5 + 0.05 **3.** ans. may vary 8/10 **4.** 80 000 **5.** 71
Wed. **1.** b **2.** none **3.** b **4.** obtuse or reflex **5.** ans. will vary slightly
Thurs. **1.** 9 000 000 cm **2.** 71 min 20 s **3.** 300 mm^3 **4.** 40 decades **5.** cm
Fri. **1.** 60 **2.** $480 **3.** 15 **4.** 12
Brain Stretch **1.** $39.60 **2.** yes, $0.40

Week 25

Mon. **1.** start at 2, multiply by 10 **2.** 67 **3.** 27, 32, 38 **4.** 61 **5.** growing
Tues. **1.** one thousand eleven **2.** < **3.** 5010 **4.** 3 **5.** 3104.97
Wed. **1.** right angle **2.** an angle less than 90° **3.** (smiley) or (smiley) **4.** 180° **5.** 10
Thurs. **1.** 8 hrs 10 min **2.** minutes or hours **3.** 6.4 cm **4.** 208 weeks **5.** 1825 days
Fri. **1.** red **2.** green **3.** 1/8 **4.** 1/4 **5.** 1/4
Brain Stretch 36.4 km in 7 day week or 26 km in a 5 day school week

Week 26

Mon. 1. c 2. ans. will vary 3. 57 4. start at 5, multiply by 5 5. 4, 7, 13, 25
Tues. 1. 12 2. ans. may vary 2/20 3. 0.045 4. 220, 202, 200, 22 5. $38.30
Wed. 1. ans. will vary ie.octagon 2. an angle less than 90° 3. circle 4. an angle more than 90° 5. 20 °, isosceles
Thurs. 1. 7:30 a.m. 2. measuring cup 3. mL 4. 3:13:25 5. 0.078 km
Fri. 1. ans. will vary 2. Leonard 3. Mary 4. 33 5. 78 s
Brain Strech: a. 1,2,3,4,6,8,9,12,18, 24,36,72 b. 1, 3, 5,15

Week 27

Mon. 1. 65, 62, 59, 56 2. 32 3. start at 90 000, divide by 10 4. 110, 132, 154 5. b
Tues. 1. 1, 5, 7, 35 2. < 3. 80 000 + 9000 + 200 + 60 + 6 4. 2845 5. 21
Wed. 1. equilateral triangle 60°, 60°, 60° 2. none 3. measure to 3cm 4. scalene 5. slide, flip or turn
Thurs. 1. 25 cm 2. 0.0009 3. 1x36, 2x18, 3x12, 4x9 4. litres 5. 144 hours
Fri. 1. not likely 2. will vary 3. very likely 4. not likely 5. likely
Brain Stretch   a. $17.00 b. $23.00 c. ans. will vary d. $30.00

Week 28

Mon. 1. 35 2. start at 44, add 1, then add 2, add 3… 3. 98, 76, 98 4. 704, 724, 744, 764 5. 84km
Tues. 1. 7000 2. 50 219.2 3. $7.65 4. ans. will vary 2/4 5. 5678.5
Wed. 1. obtuse 2. 12 3. ans. will vary 4. see picture 5. both have 4 equal sides, parallelograms
Thurs. 1. 69 hrs 20 min 2. scale 3. 0.0025 4. 48m 5. minutes
Fri. 1. 6A, 8A, 8C 2. 6B 3. 40 4. 44.3 5. 40   Brain Stretch $245.70

Week 29

Mon. 1. 100, 114, 128, 142 2. 78 3. start at 701, add 10 4. trapezoid 5. 121212, 1212121, 12121212
Tues. 1. = 2. twenty-two thousand one hundred ninety 3. 279 4. 4160 5. ans. will vary 2/6
Wed. 1. see picture 2. 8 3. all side have equal lengths 4. pentagonal prism 5. see picture
Thurs. 1. thermometer 2. calendar days or months 3. 0.0049 4. ½ b X h b=8 h=2 5. 6:12:30
Fri. 1. $36 2. $302 3. $305 4. $314 5. ans. will vary coldest month
Brain Stretch   a. 300 donuts b. $0.08

Week 30

Mon. 1. start at 499, add 99 2. shrinking 3. 50 4. 505, 456 5. 1000, 500, 250, 125
Tues. 1. 1265 2. 0.0289 3. 114 cm 4. 1.7, 7.1, 7.2, 7.7 5. 90 000 + 9000 + 800 + 7
Wed. 1. flip 2. see picture 3. 6 4. ans. will vary slightly // 5. ans.will vary- number of sides etc
Thurs. 1. 315 min or 5 hrs 15 min 2. 100 years 3. mg 4. 6 700 000 mg 5. 1919
Fri. 1. total the data set and divide by the number of values in the set
   2. middle of values when the data is arranged in numerical order
   3. difference between the highest and lowest values in a set of data
   4. a chart that uses stroke marks to record frequency
   5. x and y axis   Brain Stretch   a. 13 cases b. $375.00 c. 2 - 1 loonie 1 quarter

Printed in the USA
CPSIA information can be obtained
at www.ICGtesting.com
LVHW080322111023
760786LV00013B/533

9 781771 055604